PLANT
GROW
HARVEST
REPEAT

PLANT
GROW
HARVEST
REPEAT

Grow a Bounty of Vegetables,
Fruits, and Flowers by Mastering
the Art of Succession Planting

MEG MCANDREWS COWDEN

Timber Press
Portland, Oregon

Published in 2022 by Timber Press, Inc.

The Haseltine Building
133 S.W. Second Avenue, Suite 450
Portland, Oregon 97204-3527
timberpress.com

Printed in China

Cover and interior design by Corinna Scott
Illustrations on pages 79 and 94 by Ashlie Blake

ISBN 978-1-64326-061-7

Catalog records for this book are available from the Library
of Congress and the British Library.

For my grandmother, Anne (1900–1948),
whom I've never met but always known,
and her son, my dad (1938–2021),
whose eternal love and support is my
most treasured garden perennial.

CONTENTS

Preface   8

Succession Across the Landscape   11

Mastering Succession   25

Edible Perennials in the Food Garden   55

Vegetable Successions in the Food Garden   85

Flower Power   129

Tending the Soil   145

Seed Starting and Garden Planning   155

Spring Ahead: Hastening the Growing Season   179

Annual Flower Successions in the Food Garden   199

The Art of Interplanting   223

Garden Renewal: Succession Planting in Summer   243

The Fall Garden and Extending the Harvest   263

Glossary   274
Resources   276
Acknowledgments   278
Index   280

My hope and dream in writing this book is to encourage other gardeners to take inspiration from nature: to slow down and bear witness to the intricacies of the plants that surround us; to consider what you have not thought possible for your growing climate; and to continue to stretch yourself, just as your plants do so earnestly, every chance they get.

I have a background in natural resources, and I was trained in college to observe patterns in nature. That stuck with me as I learned gardening, fueled by a passion I shared with my husband. The more we gardened, the more the lessons of ecology became apparent, even in our annual vegetable garden.

I acknowledge that the patterns I share here and the ideas spurred by those patterns are thanks to the wisdom and land management practices of Indigenous peoples across our beautiful continent. The original environmentalists, their hands shaped the natural systems we know and love, and reference, such as old growth forests, native prairies, and oak savannas.

In this book, I often use the term "succession garden" to describe how a succession planting mindset applies to all aspects of the home garden.

A succession garden translates all facets of succession planting to every layer of the landscape. It is quite the opposite of the plant-it-all-at-once garden, and this may delight or frighten you.

Personally, I take heart knowing every single day is a day to begin anew in the garden, and a succession garden is planted over and over again, in little rows, in single transplants, tucking new life in as space allows—and sometimes demands. For those who enjoy taking their time, the succession garden is a faithful companion, always by your side. And once you have learned what to add and when, it offers endless entry points to the growing season.

For over two decades, growing food has gradually evolved from a lightweight hobby to a full-blown lifestyle and full-time job. As with the most influential and instructive life experiences, its impact on my family grew with each passing season, taking root literally and figuratively. We began our gardening careers pre-internet, relying extensively on a few key references that set the foundation for the organic practices we use to this day.

These pages, images, and ideas come to you from our brisk and bountiful USDA hardiness zone 4a garden in Minnesota. Gardening in a cold, abbreviated climate necessitates ingenuity, and we have grown to delight in the challenges of our short growing season, stretching it out as far as possible in every direction. With the ground frozen solid for more than one-third of the year, we make haste harvesting food fresh from our garden for at least eight months, from April through late November, sometime before the leeks become suspended in time.

With every growing season, we push, extend, and trial new ways of growing that fuel our joy. Failure is the most expeditious instructor, and the garden is no exception. When we succeed, it's not as urgent to reflect on why something worked well, but when we fail, it's a commanding invitation to ponder the variables that led to the mishap. Being perpetual students of the garden in this way is our deepest joy. I hope you will take some of the ideas here and apply them to your gardening season. Knowing what's possible for my family in our growing climate, I am confident everyone can feed themselves for longer than they realize. If each of our gardens stretched the harvest a few weeks each year, imagine how much more food we could collectively produce.

My wish is for this book to meet you in your garden and your life exactly where you are, and that my writing and philosophy help you examine, reflect, and celebrate all that's good, while simultaneously exploring vast, untapped possibilities for your garden. I hope through reading this book you gain an understanding of how every corner of your landscape contributes to the succession of life and productivity; and how minute shifts can impact your land and food garden exponentially for the better, while also contributing significantly to the larger fragmented ecosystem.

Penning this book in 2020, during the COVID-19 pandemic, has brought to light so many inequities. Now more than ever, every garden counts. Now more than ever, growing your own food is a seemingly simple yet staunchly revolutionary act. I've never been more grateful for our food garden, and the food security our land provides, than I have this growing season.

Above all, I hope that you continue to grow and harvest the deep joy, gratitude, and nourishment that gardening can bring, while also tending the seeds that sing to your heart.

Our food garden in late summer, starting to lean toward fall crops. A late succession of corn is a breath of fresh air, and adds structure in August.

SUCCESSION ACROSS THE LANDSCAPE

Life cycles in the garden are rhythmic and seasonal, and life succeeds through the succession of plants. As one generation completes its lifecycle, more and often different species are waiting in the wings to take its place. This constant changing of the guard is the tried and true way of the dynamic plant kingdom, and it blesses us with life: the air we breathe, shelter we seek, clothes that protect us, and food on our plates.

In the simplest terms, succession means one follows another. In ecology, succession describes how plant communities develop over time, with different groups of plants succeeding one another. Plant succession encompasses a procession of life, starting with pioneering grass species that fill in bare earth, and eventually give way to trees that, after millennia, become a beloved forest. Succession occurs in every landscape—even in our home gardens. The only difference is the time horizon we, as humans, inhabit.

Examining how various ecosystems succeed one another across generations is a beautiful way to study how we can meaningfully bring succession into our gardens. The food garden, that quintessential summer tradition, completes its life cycle in a matter of months, but the principles of succession are very much alive even in this abbreviated setting.

Summer is the most robust season in a prairie, offering the widest range of flower types that attract the broadest insect diversity to our landscape.

What Is Succession?

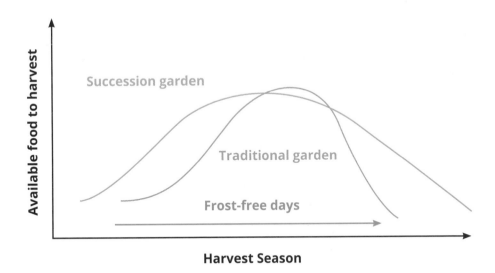

A succession garden incorporates lessons from plant communities into its very fabric. It dynamically emulates natural transitions across time. This garden nourishes you, the gardener, for as long as is feasible, because you have utilized plant diversity of both perennials and annuals to weather seasonal challenges with ease.

Succession gardening will increase your garden's productivity by maximizing the days of your growing season, even if those days are limited, and maximizing your space, even if your garden is small. It is a mindset meant to engage your imagination and your unique goals for your family's table, and fuel your stamina for sowing throughout the growing season.

The forest is an incredible teacher for the succession gardener. Though its time horizon spans human generations, it is nimble and ready for change. Armed with a deep seed bed, a forest can respond as needed to changing light levels and disease pressure, adapting and modifying its plant communities to match its dynamic environment. As you gather your seed packets for the season, so too does the forest keep a stash of seed on hand.

The once expansive tallgrass prairie is perhaps the landscape from which I have learned the most in succession gardening. The prairie is like a perfectly

A succession garden lengthens the season on both ends, and if planned well, even reduces the summer overwhelm so common in traditionally planted gardens.

orchestrated bouquet, with more than one flower in bloom each day of the growing season, always offering food for wildlife. Imagine if your garden offered you nourishment from very early in spring to well beyond first frost, beyond the traditional growing season. Again, this is my ideal garden, where copious vegetables and fruits are accessible as many days of the growing season as possible.

The common bond between your garden and these vast ecosystems is that they all begin with bare ground. Many of the same principles present in larger ecosystems abound in our little home gardens, principles such as edge and gap dynamics, overstory-understory communities, and interplanting relationships.

Lessons from the Prairie

The prairie is an incredibly dynamic and sophisticated ecosystem. It offers nectar, pollen, and habitat to innumerable insects and wildlife, and the constancy of its nectar throughout the growing season is steeped in wisdom. The lesson inherent in the dynamic prairie is that each season is enough. Flowering plants

A midsummer prairie is dotted with early season flowers like white false indigo, Ohio spiderwort, and purple meadow rue, a harbinger of the feast to come.

shift and negotiate space as the system evolves. It doesn't rely on just a few plants to nourish, but as many as are willing and able to thrive. I can't help but think of a diverse food garden when observing the prairie's cornucopia of flowers across the seasons.

Prairies, like forests, go through many stages of development. Short-lived perennials arrive early and anchor the space, while other, later succession species don't establish until five to ten years after planting. While all the seeds are sown at the same time, their maturation varies widely—a nod to the succession garden, to be sure. While the garden may be planted just once in a season, it is filled with vegetables that vary in their days to maturity, and so the harvests extend across several months. While an orchard is planted once, its maturation also spans several years.

Constancy is the prairie's foundation. It times its flowering to be seasonally constant, as the insects it relies on for pollination have evolved in tandem with its rhythms to successfully produce the seeds of life that ensure it flourishes. I am simply in awe of the sleek yet understated sophistication of the mature prairie.

Like a well-planned succession garden, the prairie offers modestly in the shoulder seasons and more robustly in summer. But the prairie's real trick is that its offerings during spring and fall are sufficient to feed wildlife in the

The revolving door in the early years of a planted prairie: early Canada wild rye fades to yellow and the coneflower starts to sunset for the season as stiff and early goldenrod arrive.

growing season's marginal weeks, and the same holds true for how the succession garden can provide for you. The prairie's steadfastness as a source of nutrition for its communities, even in the lean months, reminds me there is always a way. The key for the garden is to bring the right foods in at the right time.

In this way, the prairie teaches us that abundance need not always equate to excess—that, in fact, abundance can be simply having enough; that what is produced and offered each season is sufficient. This is the rhythm by which seasons ebb and flow in a naturally interplanted prairie: a little bit is offered, for an extended period of time, nourishing at a more sustainable rate than the predictably overwhelming harvests of summer.

The height of summer is a naturally overflowing moment in the growing season, even in the prairie. There, summer's floral explosion is commensurate with gloriously long days and ample rain, and the prairie offers robustly during this midseason floral succession. Similarly, the food garden offers the largest cornucopia of variety during the hottest weeks of the year. For the prairie, however, the season of abundance stretches well beyond summer's floral fireworks.

I don't know about you, but I prefer a steady, manageable stream of food from my garden all season long. It seems most years, during the height of summer, food comes at us in waves and quickly seems insurmountable, a veritable burden of abundance. The ingenuity of the prairie is how it spreads the feast out and marvelously times its blooms for as long as possible. And the makeup of the prairie ebbs and flows across the years—some plants are short-lived, while others arrive later, heralding maturity, inviting us to consider how that can apply to our food gardens.

I'm here to tell you that with a succession mindset, your garden can feed you longer too, taking its cue from the prairie. Let's consider what that looks like. What if you actively tended growing vegetables under a low tunnel while late winter and early spring frosts persisted? What if you timed your plantings so the harvests arrived across an expanded time horizon, instead of just a few frenzied months? Could this possibly meet your food needs, reduce seasonal overwhelm, create a more robust local food system, and thus feed you more completely?

As the growing season fades against the cold of late autumn, what remains in the prairie is a substantial bed of seed, a veritable grocery store for flocks of hungry birds and small mammals. Likewise, you have the opportunity to stretch your harvests by growing vegetables that persist beyond the first killing frosts.

You can put up shelf-stable vegetables in root cellars and store dried goods in your pantry. If your climate is mild, you can tuck root crops under row cover or mulch for winter harvesting, or grow overwintering crops like sprouting broccoli. Another beautiful lesson offered by the prairie is how to embrace these lean, dormant months of winter by producing and gathering a steady stream of nourishment for them.

Lessons from the Forest

The forest offers wisdom for the keen observer in many of the same ways prairies do. The forest imparts lessons about light availability; the relationship between overstory and understory plants; disease management; and the merits of planting diversely, all with whispered caution of the perils of overplanting. I see vegetable gardens as little fairy forests whose maturity, while it happens at warp speed relative to a forest, mimics much of what we understand from studying forest dynamics.

For most, more light means happier, stronger plants. Sure, there are shade-tolerant species, but they're largely the exception, particularly in food

At a foot tall, a closed canopy of garbanzo bean plants perfectly demonstrates the relationship between plant spacing and light availability.

gardens. Plants that suffer from low light do not develop as strongly as those with adequate light, predisposing them to all sorts of challenges.

In a dense, mature forest, it's easy to observe the relationship between available light and canopy structure. Light penetrates as deeply as possible, but when a forest canopy is closed, most of the green is way above your head, consuming most of that light. A closed canopy is filled with mature trees, their branches stretching out to meet one another some tens of feet high, occupying every last square inch available in which to produce energy and thrive. (The space their leaves occupy is called leaf area.) At ground level, there is very little understory because of the lack of light that reaches the forest floor.

Your garden creates canopies of many different heights, depending on the crops you grow. My rule of thumb is that I want the plant canopies, no matter the crop, to remain in full sun and grow openly until they're more than halfway to maturity. So, eventually, I expect my tomatoes to mass together in a big, beautiful hedge, but not until they're at least several feet tall. The space you allot aboveground to your plants is more or less equivalent to the space they occupy belowground. And while it is possible to give plants too much space, more often than not, the opposite occurs.

When plants are young, sunlight likely reaches all the way to the ground, regardless of how densely a garden is planted. How long it remains that way is a

Pushing my luck with this interplanting, the kohlrabi and onions spaced too close together, the first row of onions stunted from lack of space and light. Kohlrabi's massive leaf area shaded the onions out.

function of who's growing, their initial spacing, how fast they mature, and available daylight hours. Capitalizing on all this sunlight is both feasible and advisable. You can grow more than one crop in a space when your dedicated overstory plant (the one that will remain in place longest) is young. Quick succession crops are your understory, and before or as the canopy closes, these understory plants will be mature and ready to harvest.

Ever notice spindly understory trees trying their hardest to reach for distant sunlight as it's consumed by the overstory? Perhaps they remind you of struggling vegetables who may have been crammed together because you simply had to grow them all? Older, mature forests with multiple canopy heights teach of interplanting as a viable technique for gardens, used with caution. Considering the maturation rates of plants paired together is part of the equation, as well as ensuring appropriate light levels remain throughout the duration of each plant's lifecycle.

Walking through the woods, have you ever taken notice that suddenly you're in a sunnier spot? And in that gap in the forest, the composition of plants differs from where you just were? These gaps provide new life, and a chance for the seed bed to shake things up, bringing diversity to the ecosystem, all thanks to a tree that fell and opened a hole in the canopy. Depending on what plants establish in these gaps, the species composition will grow and mature at a different

A July gap in the cabbage bed created by harvesting the last of our interplanted beets, arugula, and spinach opened space for summer sowings of carrots that will germinate under damp burlap.

rate than its surroundings. These plants offer themselves to the landscape at different intervals, creating a few successions within the same space. Gaps like these create extended harvests that are precisely what we are aiming for in our home garden. Has your perfectly planted garden ever been menaced by an unwelcome disease, a hailstorm, or a garden pest? I see these calamities as opportunities. Of course, they come with a heavy dose of humility, but the gaps created by these mishaps (abiotic and biotic) immediately spark a new growing season. What you choose to do with this extra season is up to you, your creativity, your growing climate, and your seed stash or garden center.

Most forests hedge their bets against these inevitable bumps in the road with diversity. So should every gardener. The more diversity you add to your garden, the higher the likelihood of your success. Those unplanned gaps become smaller and smaller as you spread your garden out across the growing season and across plant groups. You can hedge your bets by cultivating a robust list of fruits and vegetables that excel in every pocket of each growing season, again, to create constancy (a trait the forest and prairie share).

I marvel at forest edges—the space where meadows, fields, or bodies of water meet the forest. The forest edge is naturally sunnier than its companion forest. More light penetrates the edge, decreasing precipitously as you head deeper into the forest and the canopy closes. Edges typically display a bouquet of green from forest floor all the way to the top of the canopy. Edges thrive with diverse plant growth at nearly every height. It's a beautifully orchestrated display of plants negotiating space.

Your food garden is full of edges. An edge borders every raised bed. Some edges face the sunny south side, while others meet shadier northern aspects. As with forest edges, these unique spots within your garden are where light penetrates the full height of the bed, despite plant competition, and they're also prime opportunities to interplant. Not only are edges eye-catching and diverse, they are productive. In raised beds, edges offer opportunities for flowers, herbs, and vegetables to cascade. You can interplant in these areas with more success than under or near overstory plants. My garden edges are often dedicated to pollinator or salad garden habitat.

Edges have their limits as well. Even a full sun edge can be planted too densely, and a rush to establish dominance ensues. Proper plant spacing will ensure the plants you add to your understory and along your edge have room to thrive, creating their own closed canopy as they reach maturity, adding a second canopy to the planting, and visual, edible interest to your vegetable beds.

Oak Savannas

Autumn gold in a restored oak savanna prairie ecosystem in Minnesota, where the dominant landscape remains in full sun.

Oak savanna is a blend of prairie and woods, overstory trees dotted within thriving open spaces. Leaning more toward prairie, this ecosystem is characterized by low density, open-grown trees. A diverse community of herbaceous plants flourishes in the vast areas of open space between the trees, like prairie grasses and flowers. As the trees grow tall, establishing themselves between fire cycles, they persist across the decades, while the understory plants ebb and flow.

In the span of a growing season, you can create these sweet little pockets within your garden through staggered successions and interplanting. There's something very grounding about the open-grown sturdiness of an old oak tree, and my autumn kale "trees" are that for me. I set our earliest kale in place in late March, as tiny little seedlings. They grow up among a cohort of early producers, and the canopy of brassicas closes by early May. Soon after, we harvest the surrounding food, leaving the kale plants in place to dominate the bed for the remainder of the growing season. With a blank understory, we sprinkle carrot seeds, a perfect low-growing companion to our open-grown greens. As the kale plants mature and gain stature as foundation plants in our annual summer garden, the understory of carrots takes root, creating a mini, edible oak savanna.

Mixed Species Forests

Older succession forests with mixed species are some of the most resilient of all. These forests have repopulated gaps at various intervals over time. This uneven-aged canopy is the portrait of species diversity and stability. When calamity blows through, the mixed forest is protected thanks to its species composition and age diversity.

The mixed forest is the ultimate interplanted forest. It includes several different tree species growing together. This ecosystem is the antithesis of modern agriculture, where the orderliness of monoculture (a planting of a single tree or other plant species) eases overall management at the cost of biodiversity and resilience. Unlike a managed, even-aged forest, the mixed species forest is a living tapestry. With canopy diversity, this system thrives as light penetrates

A small seedling in late March, Scarlet kale now anchors this late summer bed and creates little trees amid a bed of carrots thriving in the understory.

to different heights throughout. Each species has different nutrient, light, and moisture needs, and thus the system is more resilient and adept at weathering annual climatic variabilities.

The home garden often embraces order because of its inherent ease of management, and while growing large blocks of the same thing does simplify matters, it is not the way of the natural world. Implementing smart and thoughtful interplanting, on the other hand, will improve your garden's resilience, increase productivity, and foster healthy soil. The natural world's system is tried and true, and embracing diversity is what we are called to do in this moment.

What natural ecosystems speak to your own gardening style? Where do you derive inspiration for your garden? Let's dig deeper into how you can embrace the lessons of the prairie and the forest to bring food to your table for weeks or months more, across the four seasons.

A diversely planted garden is both aesthetically pleasing and practical, offering smaller gaps to replant throughout the season.

The garden is always between seasons, and yet always in season. We are present in one season while preparing for the next, enjoying the moment while implementing an actionable plan for the months ahead.

MASTERING SUCCESSION

The foundation of all succession plans is just that: a plan, and a thorough one. In order to maximize joy and productivity across the four seasons, your plan requires a grounded presence in today, as well as clear foresight of impending seasonal shifts, no matter the size of your garden. With a keen eye to the present, the fruitful gardener can embrace the now and derive just as much joy in what is to come, cultivating abundance with the seeds of today.

In this chapter, I will discuss strategies for succession planting and how to implement them your garden, as well as some dos and don'ts. Those strategies include:

- Continuous planting
- Variety
- Blocking
- Endurance
- Zone bending
- Interplanting
- Staggering the harvest
- Food with flowers
- Vertical gardening

Balancing presence in the moment while dreaming, scheming, and taking action toward the season ahead is the basis of succession gardening. Gardens are dynamic systems. In the case of food gardens, the passage of time is marked by a near constant arrival and departure of crops as seasons blend, overlap, and play together across weeks, and in milder years and climates, months. Anticipating gaps in your garden, selecting plants that will produce over an extended period, and keeping the garden renewed with transplants are all excellent strategies.

Succession planting is the act of maximizing your garden. By amplifying what nature has taught us, my family has cultivated a garden that is exponentially more productive. We have learned to maximize our growing days, producing as much as we can on each square foot of soil in the shortest possible timeframe. You can accomplish this in your garden too, in a multitude of ways. Each approach has many benefits, and one or more can and should be used in conjunction with another to create a dynamic garden landscape. Let's discuss the most common ways home gardeners and farmers alike achieve continued and extended harvests.

Continuous Planting:
The Art of the Now and Then

A tool for all climates, continuous planting turns a typical "main season" garden into a spring, summer, fall, and winter garden. It transforms the garden into the produce aisle, providing food in every season. Executed well, continuous planting produces a diverse array of food throughout the growing season. The garden becomes a place where spring cabbages are met with early summer cherry tomatoes, and then late-ripening peppers collide with fall apples and frost-kissed Brussels sprouts. Continuous planting is the ultimate garden tool, enticing you to experiment and try something new, to stretch your season a week or a month longer, and to keep your sowing stamina strong.

Grasping the notion that a garden is never fully planted is key to embracing the concept of continuous planting. Think about how florists work to get the most vase life out of their bouquets by knowing which stems will last longer and which will need to be swapped out. The succession gardener must implement continuous planting in a similar way. Sure, your garden would be a satisfactory

Just as forests and prairies progress through time, the annual food garden has a rhythm and succession as well, with a time scale measured in one growing season rather than years or decades.

afternoon project on a sunny, late spring weekend, planted in a day and done. But that also means yields will be concentrated into a narrower time horizon, and harvests more meager than if you planted over a longer period and more diversely. Even just a once-a-month reminder to sow something for the next season is a major shift in rethinking the garden as the revolving door it truly is. The concept of continuous planting is fluid. I think of it like a garden renewal; motivation to keep an eye on your space, what's maturing soon, and where space will open up for new plantings. Having a plan for the food or flowers you want to tuck in or direct sow as soon as that space is free makes the process more approachable.

Continuous planting is its own kind of garden plan. The clearer you can be with the types of food you want from your garden, the stronger your plan will be, and the more motivated you will be to implement it. Do you want salads all summer long? Get your pen and paper, scour the seed catalogs for heat-tolerant lettuce varieties (there are many to choose from these days), and sow a diversity of lettuce from late winter all the way until the end of summer. How about green beans three months straight? There are varieties for that. The key is to embrace seed starting, which opens up a truly endless world of possibility for the curious and ambitious grower, with new varieties released annually.

While seed starting (direct sowing) in the early season is easier—cooler temperatures, fewer distractions, more consistent moisture—plants establish much faster in the warmer months. So, while it feels more arduous to sow beets in the heat of early July, they take off more quickly than those transplanted under row cover in late March. Tune in to the intricacies of your growing season and the cravings of your palate to work through these details.

Continuous planting is the heart of the succession garden. If the thought of it sounds, well, exhausting, that's because it is. I'll be the first to admit it. Truth be told, it is a practice in mindfulness, of being aware of your garden's needs now, and in a month, and next season. Enjoying where you're at and seeing where you can be. Tuning in to where and when space will materialize and continuing to renew that space, maximizing the productivity of your garden. When you succeed at feeding yourself, your family, neighbors, and friends, the rewards more than compensate for the fortitude you had to muster to keep sowing.

Started indoors in early July and transplanted in early August, these heat tolerant head lettuces settled right in after our garlic harvest at the end of July.

Variety: Single Crop with Staggered Maturities

It's time to learn a little secret: you are already a succession gardener. The cornucopia of food that rolls out of your home garden each growing season, collectively, is a series of successions. Let's take a moment to appreciate that even a garden that's only planted once and subsequently tended, watered, weeded, and enjoyed is its own marvelous parade of successions. Now let's break that cornucopia down and say that instead, you plant variations of a single crop repeatedly, all season long, and reap a steadier series of harvests over a longer period. This method illustrates the succession garden concept of staggered maturities.

Planting a few different varieties of the same vegetable (or fruit, for that matter) with various maturation dates is the simplest way to implement staggered succession planting. This takes a bit of research, but the process of researching new varieties is a lesson in the diversity of seed for every vegetable—and a chance to marvel at how much variety is available beyond grocery store produce, hidden treasures known and loved by the home gardener.

The best example of how a single succession can become a staggered harvest is the beloved garden tomato. Tomatoes come in endless shapes, sizes, colors, and textures—and permutations therein. The tomato is a well-loved and tended string of plant DNA, played with so much the diversity offered to gardeners and consumers today is practically boundless. From carefully preserved heirloom

A well-planted edible garden is the ultimate feast. Similar to a forest or prairie in that it provides nectar and food across the four seasons, a well-planned and planted garden makes the most of the space and time.

varieties, to those newly discovered, to disease-resistant hybrids, there's a tomato for everyone and every garden.

When you plant one each of cherry, beefsteak, and paste tomato in late spring, you will have a steady procession of tomatoes all summer long, and thus a staggered succession. I know tomato gardening no other way—my cherry and beefsteak tomatoes always ripen at different times. The diminutive cherries kick off the summer feast, their small fruit maturing more quickly. Beefsteaks take their sweet time, slowly accumulating enough sunlight and energy to produce one-plus pound fruits that arrive mid- to late summer. The paste tomato (often a determinate variety) comes in with a crash and hangs on for a several weeks, commanding late nights of canning for the promise of a warm winter's meal.

Peppers are another plant-it-once vegetable for reaping extended benefits. Similar to tomatoes, in general, the smaller the pepper, the faster to maturity. I often see color on my little hot peppers before the thick-walled bell peppers hint at the slightest blush of anything but green. So when you're dreaming and

Even a sowing of the same variety will not result in maturity at the same time. I would not be nearly as excited or enthusiastic about my broccoli planting if all five of my plants matured within two or three days.

scheming your pepper garden, be sure to include a diverse lineup of pepper types of various sizes to stagger out the harvests of this summer darling.

One of the easiest foods to direct sow is beans. Pole beans thrive as a single succession that just keeps on giving, producing a consistent harvest indeterminately, making them an indispensable food to grow if you want to feed yourself seasonally in the warmer months and don't want to replant your garden. Bush beans, on the other hand, have a narrower harvest window; they will grace your table for several weeks, and benefit from continuous planting. Plant some of each, and you'll have enough beans for your meals with plenty to share.

Regardless of the size of your garden, mixing and matching seed of the same vegetable extends the life of any given crop while coloring your plate with a whirlwind of variety sure to please the crowds. In this way, a single crop becomes its own succession planting.

We implement staggered maturities in every food we grow, at least two types of each, naturally extending the season as much as a month or two. This also makes for more interesting meals all season long.

Blocking: Back-to-Back Plantings

There's a place for wild gardens, and there's a place for order. Blocking is a succession planting technique that leans heavily on order, achieved and maintained by a very active garden hand. Blocking is tidy rows and evenly spaced transplants or sowings of the same plant, reminiscent of those vistas of orderly rows of rural cornfields on long road trips. Block plantings like these are beautiful in their scale. However, I believe there's something intrinsically appealing about this modular approach in a garden of any size.

Organic farmers utilize crop rotation to build soil structure and renew nutrients. Block planting in large areas makes for easy management of cover crops such as these peas.

Many of our agricultural crops are harvested and planted as single successions, from corn, soy, wheat, and sunflowers, all the way up to the coniferous forests of the Pacific Northwest, grown for timber. This is a practical method for growing food too. Since the advent of industrialized agriculture, massive tracts of monocultures grow at more or less the same rate during the hot summer months across thousands of miles. Ease of management and profit are the driving force behind this approach. This comes with stark environmental tradeoffs: biodiversity is reduced above- and belowground; habitat is eliminated for

Onions are one my favorite crops to plant in dedicated blocks. I replant the space with a fall crop mid- to late summer as a second succession.

native insects and wildlife; and critical migratory habitat for birds is reduced or eliminated. In spite of the harm, though, there are lessons to be learned from industrial farming.

I first explored gardening using back-to-back plantings, sectioning out planting blocks, allocating space for one food per area at any given time. It made managing the garden very straightforward. When my broccoli was all harvested, I'd sow a succession of late summer green beans. After our earliest carrots were harvested, we would sow fall snow and snap peas in the gap. I'd prepare our fall cabbage bed with a buckwheat cover crop in May. The buckwheat would mature just in time to cut down and transplant the fall brassicas. If I needed to weed around my garlic, it was easy to access and tidy up, having been planted in straight rows; once it was harvested, some fall lettuce, radish, and celery took root. Even on a small scale, planting a space that has recently been harvested and cleared is much easier than negotiating a space with some plants still maturing.

Blocking is a method I continue to implement in my home garden for many food groups. This is a bit at odds with interplanting, a succession technique you will learn more about later in this chapter—as fun and interesting as it would

Planting blocks vary in size depending on the vegetable. Here, our AmaRosa potatoes are the only thing planted in this 4 ft. by 10 ft. stretch of our in-ground beds, for ease of hilling.

be to interplant everywhere, sometimes practicality outweighs the whimsical beauty of garden chaos. And, personally, I welcome an orderly portion of the garden. It becomes like art, definitive and understated, and practical. Very practical. Less is sometimes more.

Some foods are better suited to their own space, rather than mingling with neighbors, and blocking works well for them. Maybe they don't grow very tall, thus faring poorly with light competition. Perhaps they must be mounded midseason to produce tubers (as with potatoes) or ground nuts (as in peanuts), and disturbing nearby soil would disrupt or damage interplanted neighbors. Planting in tidy, micro-monoculture rows also eases weeding and overall management, and that's the reason market gardeners and farmers of all scales implement this approach with reliable results.

The longer I garden, the more my approach has evolved, and the more I appreciate the merits of each type of succession planting and apply them methodically throughout the garden—even blocking, in spite of its nod to industrialization. The beauty of gardening is that it holds space for tidy rows and wild experiments alike.

Endurance: Succession with Continuous Seed Starting

Cliché but true, timing is everything, and that includes the garden, especially when you are faced with a short growing season. Whether it's dry season in the high desert, frigid winter in northern climates, or sweltering summer heat in the subtropics, time often works against the gardener. What better challenge than to try to outwit it? That's precisely where the succession garden triumphs.

The ultimate garden goal is for all your beds to actively grow and produce for as long as possible throughout the year. To that end, maintaining your very own little nursery of seedlings is a prescient strategy to prepare for the inevitable, such as prolonged heat waves, intense pest pressure, and early frost. I did not say this strategy was easy. (Spoiler: it's not.)

Seed starting is a marathon, not a sprint. A strategic lineup of veggies every few weeks helps us maintain a highly productive home garden.

Dividing and transplanting direct-sown kohlrabi seedlings in a late summer gap extends our fall harvests.

Keeping a continuous supply of vegetable starts at the ready takes planning. I plan a month or more ahead, adjusting my sowing schedule in real time each year as I watch my plants grow, looking for veggies that are maturing too slowly or bolting too early. This informs my month to month sowing needs. If I'm feeling too busy to sow trays indoors and there's a gap from a recently harvested crop, I can use the garden as a nursery, though this is not as efficient a use of space as starting indoors.

Sowing plants indoors gives them a chance to germinate without excessive heat or pest pressure, presuming there are not cabbage moth or flea beetle outbreaks inside your home. Providing plants with as gentle a beginning as possible allows them to develop strong root and shoot systems without pressure from environmental conditions outside your control in the garden setting. When it's time to transplant into warm garden soil, starts are sturdy and strong enough to withstand minor damage from pests.

In our intensive home garden, indoor seed starting begins in early February and continues steadily until at least mid-July. At five months of seed starting, that's as long as our outdoor growing season—and the key to our extended harvests. We begin the indoor sowing season with plants that are slow to mature, like onions, peppers, eggplants, and celery. These are followed by shoulder season stars, headlined by the brassica family; we rely heavily on cabbages and broccoli to ensure robust provisions, and sow these several times in succession all season long. Flower seedlings mingle with vegetable babies during spring sowings, and in the meantime, I direct sow a short list of quick crops outdoors.

Continuous seed starting is one of those "just do it" techniques. There's no easy way to get around the work of it. It takes a great deal of endurance to continue to sow seeds indoors when the weather warms up, the garden is already fully planted, and you're harvesting. But it is a key tenet for succession planting to come to fruition, and the clear difference between a great food garden and an extraordinary one. Sure, you can seed tomatoes directly into the ground when the soil warms, and you'll likely have enough time to enjoy bowls of cherry tomatoes before a first frost kisses the plants goodnight. But with an indoor head start, we enjoy ripe cherry tomatoes about a month after our last frost—three months earlier than those that are direct sown. With this mindset, you can sow more than one succession of tomatoes and have robust, productive plants all the way to the moment that first icy frost bids them adieu.

Zone Bending: Extraordinary Measures for Season Extending

Living in a cold climate brings to light very quickly just how important every possible growing day is for an edible garden. Our growing season comes to a close much faster than our hearts, minds, and bellies can handle, sometimes leaving us distraught with its sudden departure. My first frost often arrives on an early to mid-September evening. Making the most of your climate requires understanding and appreciating its limitations while seizing opportunities to extend the season. Zone bending offers the adventurous gardener a new take on the traditional growing season, regardless of zone.

Playing with the seasonal shoulders of our favorite food crops helps us extend our growing season. Sweet corn can be planted anytime over a two-month period. Sometimes we sow in late April, other times in late June.

My gardening philosophy took root while I lived in a significantly warmer zone, so I am intimately aware of the flexibility a longer growing season affords the gardener. In warmer zones (zones 7 and above), you can take your time getting hot season crops planted because there's a much longer season for the plants to establish and mature. That said, it's important in warmer zones to learn to maneuver around the pounding heat and pathogen-prone humidity of summer. The answer is to plant earlier—and plant again, later—than you normally would. There are several workarounds to the perennial challenge of your individual climate, wherever you live.

In northern climates with extended cool periods well into meteorological spring, erecting low tunnels is what I've found to be the most emotionally uplifting and practical method of zone bending. Simple enclosures of many kinds have been used for centuries to create microclimates in the garden and extend the season. Predating the seventeenth century, the orangerie, for example, began as a makeshift enclosure to protect frost-sensitive orange trees. The idea grew into an ornate room, and remains a sign of prosperity even today. This method remains tried and true for zone bending, and low tunnels are the

Springing our growing season ahead with simple low tunnel enclosures earns us a six-week head start on the growing season.

economical and effective—and tremendously efficient—modern-day equivalent. We consistently harvest early, fast-growing crops like radishes, lettuce, kohlrabi, bok choy, arugula, spinach, and green onions before our last frost in early May. Most gardeners here haven't even planted their gardens at that point, which shows just how much room there is for growth in the succession garden.

Many seasoned gardeners believe that the growing season commences after the chance of last frost has passed. While that's definitely true for some extremely popular main season crops intolerant of frost, on the whole, this out-of-date belief unnecessarily abbreviates the gardening season. There are means to circumvent this (low tunnels, for example) that put the garden well on its way to being a self-sufficient grocery store by that same benchmark in spring. And the rewards are profound.

While our frost-free growing season is 4½ months on average in Minnesota, we consume some form of homegrown food all year long. We don't have a greenhouse, but do grow things that store well and feed us from fall through spring, from garlic to onions; winter squash to potatoes; Brussels sprouts, fermented napa cabbage (kimchi), and canned tomatoes.

The majority of our zone bending leans toward spring. It's the time of year when my energy for outdoor work is fresh, when the days stretch longer and warmer, and when working long hours feels appropriate and rejuvenating. And I am convinced I am merely following the plants' drive at this time of year to grow faster and faster with the rapidly lengthening days and warming soil. Their exponential spring growth meets our plates and hearts with reverence.

Interplanting:
The Ultimate Succession Tool

The pinnacle of succession planting is interplanting. This means growing more than one type of vegetable or flower simultaneously in the same proximity. Also called intercropping, and sometimes referred to as companion planting, this mingling of more than one type of plant in a space creates a living mosaic—quite literally a garden salad.

Far and away my favorite way to plan and plant the garden, this method succeeds on all fronts. It maximizes planting space, limits weed pressure, and is the most attractive way to garden, creating a lush bed in a short amount of time.

Let's just remember who taught us this. The prairie is an instructive example of interplanting. Chaotic yet harmonized, prairies evolved to produce the most

Interplanted lettuce, sweet alyssum, and red cabbage will steadily feed us and the insects from early spring through the height of summer.

efficient and nutritious buffet for local fauna. Its parade of food (nectar, pollen, and seed) is awe-inspiring, arriving early and maintaining a diverse offering throughout the growing season. It's a fundamental food source for wildlife at key periods of time, like the late season, including native bees and migratory creatures the likes of monarch butterflies and songbirds. Our food gardens can emulate that succession through interplanting, not only feeding us, but also the insects, and thus birds and other creatures too. If that isn't a win-win, I don't know what is.

Let's be clear, though: cultivated gardens are for our enjoyment. Visual aesthetics are as important as attracting pollinators, cutting flower bouquets, and

A raucous display of interplanting, this bed was thrown together with leftovers and quickly became one of the healthiest beds (albeit difficult to harvest) in the garden during the worst drought we'd ever experienced.

consuming the delightful cornucopia we dreamed up in our hearts and executed with our hands. Why not make the vegetable garden a haven and sanctuary? There's nothing I love more than a well-executed mashup of textures and colors, flowers and vegetables.

I'm not the only one. Insects thrive in a garden where interplanted diversity reigns, finding copious options not only for nectar, but also for cover—and in the case of predatory insects, often food they depend on in the form of aphids or larvae. Our gardens should be places where our hearts and minds find solace and inspiration, and when created with a succession mindset, they become equally important resources for the foundational species of our food system.

Staggering the Harvest: Slow and Steady

As you know from staggered maturities, even if you plant your garden on the same day with all the same plants, those will not all mature on the same day. Instead, they will be harvested over many weeks, even when growing the exact same variety from the same seed source. Some seeds (and thus plants) are inherently stronger than others, so they mature faster. The succession garden aims for as long a season as possible for each of the foods you grow, keeping your plate fresh and your menu diverse. Staggering maturities by planting copious varieties of specific vegetables helps, as does interplanting, combining foods with varying maturation rates in the same space. Blending all these different maturity rates multiplies the productivity of a single space.

With the majority reaching a similar height at maturity, brassicas are happy interplanted, naturally staggering the harvest from a single planting over several weeks to a month.

My experiments have proven time and again that anything goes. Mix plants together that make you happy, or mix plants you've noticed seem happier together. For me, those pairings depend either on aesthetic nuances or are driven by a plant's space needs. I subscribe to the "Does it work for me?" mindset. The key is

Four weeks makes plain just how much room a squash will quickly consume.

OPPOSITE
The perfect pair: a vegetable garden rung with flowers attracts pollinators and makes the garden a gathering space for all.

to think about the spatial dynamics of the bed, how each plant will grow, and how they will eventually mingle. Low-growing plants should be at the front of the bed, and taller plants strategically placed either in the middle or the rear.

Another way to stagger harvests is by mixing and matching homogenous crops in a planting block. I rely on this method for brassicas, a large group of vegetables that all reach similar stature at maturity. Because they grow so similarly, I know I can plant them together, appropriately spaced, and enjoy an extended harvest. Red and savoy cabbages, for example, make great neighbors. Romanesco broccoli and Violetta cauliflower comingle happily, as do bok choy and kohlrabi. Any vegetables with similar growth habits and maturation rates are an easy path to succession planting. The space becomes a veritable produce aisle for weeks on end when growth rates vary among the plants as they mature—and that undoubtedly benefits you, the consumer.

One of the most challenging things to do in the traditional summer garden is to plant a single summer squash seedling in a 4- by 4-sq. ft. area and walk away without adding anything else. All that wasted space just sitting there for a spell, all that soil underutilized. But the summer squash will overrun the space; it's not a matter of if, it's a matter of when. What's a gardener to do?

Fast-maturing veggies are interplanting friends in situations such as these: the mighty radish, leaf lettuce, spinach, arugula, kohlrabi, and even early summer carrots and beets. Not only won't these low-growing vegetables impede establishing neighbors, they will complete their lifecycle before the zucchini becomes the reliable, beautiful monster it's destined to be.

Food with Flowers:
A Match Made for Every Garden

Flowers make people happy. Food is a delight to grow, tastes more delicious
when homegrown, and sustains life. Together, they're garden magic. I've always
loved growing flowers around my vegetables, and a few key species have long
adorned my summer food beds.

Flowers have an undeniably positive impact on our lives. They dramatically improve my mood, and so it's only logical to want flowers where we spend our outdoor time. For me, that's in the vegetable garden. I want to commune with the butterflies, be distracted by bumblebees, be joyfully enticed to follow any meandering new insect visitor as it peruses our offerings, hoping to catch enough of a glimpse to later identify it to the species level. Because growing food is our priority, to immerse myself with flowers, I must immerse my vegetables within a flower garden. More than just beauty, this interplanting method attracts all kinds of insects, which pollinate at least one of every third bite of food we consume. Beyond that, it's socially and environmentally imperative to bring as much diversity of life into each of our gardens as possible.

Any cucurbit seizes the opportunity to ramble up a trellis. Here, a watermelon clings to its support structure on a hot August day.

Vertical Gardening:
Form and Function

The most seasoned garden designer, including mother nature herself, knows just how important vertical elements are in the garden. Our food garden is no stranger to well utilized vertical space. Take a look at natural landscapes and you'll quickly notice vines ascending trees, shrubs interluding in opportune gaps in otherwise dense forests, and other plants growing upward for sunlight in creative ways.

Many of the most popular homegrown foods happen to be vining crops: tomatoes, cucumbers, beans, melons, and squash. Yes, these plants can ramble horizontally on the ground and produce well, and that's exactly what they did in their wild forms as they sought more growing space. Many a home gardener and farmer alike continue to embrace the horizontal growth habit of vining crops who aren't supported. But we trellis our vining crops not just for beauty, but for control.

Trellises are pivotal for vertical interest in the vegetable garden. They also make the garden more ergonomic, bringing rambling harvests up to eye level. This additional square footage increases air flow around foliage, reducing disease pressure. Pruning the adventitious suckers off vining crops can mitigate disease and maximize productivity before it ramps up in late summer. Because many vining vegetables top out around 6 ft., adding structures to the food

garden also converts what might be a visually monotonous landscape into a more appealing and inviting space.

And then there are the crops that just grow tall, creating their own verticality, such as corn. I grow dent and flint corns as living hedges. Positioned in strategic areas of the garden, they become an edible screen, easily exceeding 10 ft. in two months' time, and readily act as trellises for vining crops like beans. Fast-growing corn is to the runner or pole bean what trees are to native vines, and practically in the blink of an eye.

A summer garden planted with corn, beans, and squash is a combination known as the "three sisters." Originating from Native American agriculture, these three crops (one vining, one rambling) planted together are a perfect, self-sufficient example of interplanting and vertical gardening. The interplanted beans run up the corn, fixing nitrogen in the soil and helping to shore up its shallow roots. Underneath, the cucurbits (squash) run amok horizontally, providing a low-growing living mulch that reduces weeds and retains moisture.

I tried this method for the first time several years ago. I searched around online for some guidance as to how to plant them together, and landed on making hills with corn in a ring, and one bean seed for each corn seed. I waited to plant the squash until a bit later. Hilariously, my corn grew much faster than the squash and shaded it out, and if that wasn't enough, in a lapse of judgment, I planted a bush-type bean, so there were no picturesque climbing bean plants. My ego was bruised, but the beans still produced a decent crop. I continue to play with interplanting this trio, always giving my corn a head start, planting beans a few weeks later, and finally transplanting winter squash starts once the beans have put on their first set of true leaves.

Diversity is the backbone of any planting, including vertical ones. Adding garden structures requires some resources, so utilize the investment to your advantage and plant as many different vining edibles on a single trellis as you feasibly can without overcrowding. Remember, diversity thwarts disease and minimizes redundancy in the kitchen. The bigger the trellis, the more space you'll have to explore interplanting combinations. The options are truly endless, and the best way to get a flavor for your own style is to dig in and go for it.

Three sisters interplanted with both flour and popping corn, three bean varieties (two pole and one bush), and one winter squash variety. Maximized diversity in a single bed, with additional vertical space for the squash to roam—and the squash always seem to take me up on that.

Space Savers: Low-Growing Foods

As much as I am passionate about adding vertical interest to the garden, it should come as no surprise I also love low-growing, compact crops too. A favorite

time of year is mid-July, when the vines are establishing but not overgrown, and all the bushy, compact crops are sitting tall and proud just a few feet off the ground. They are so orderly and well-behaved, and a stunning feature of the early summer garden.

These low-growing, well-behaved edibles belong to the non-vining food groups: alliums, brassicas, all the root crops including potatoes, plus herbs and greens. These plants top out around 2 ft. or less (excepting the potatoes), making them the tamest garden companions. Their compact growth habit provides a long garden view, contributing to and borrowing texture and form from neighboring plants.

These low-growing foods shine in mix-and-match layouts. Create geometric patterns with your food by contrasting color and texture together for an edible arrangement that is never the same two seasons in a row. I love this approach

Savoy and red cabbages mature slowly as the romaine nears maturity, with their canopies just touching. It's almost time to deconstruct this living art.

Sweet alyssum's softness edges our beds, while calendula mingles with our eggplants.

OPPOSITE
A north-facing shoreline of Burntside Lake in Minnesota is filled with a layered canopy. Sun-loving aspens thrive in this pocket of sunshine where light penetrates the entire canopy at the edge.

so much that when the time comes to harvest my interplanted lettuce that grow among my spring broccoli plants, it's a bittersweet reality to quite literally deconstruct my edible work of art. Compact plants are a fantastic use of space, especially in an urban setting. Low-growing plants naturally require much less square footage and fewer inputs than vining crops. The main difference between vining and non-vining plants is that compact plants tend to be plants that are harvested once, or over a shorter time period, and then they're done. In contrast, vining crops produce for several months at maturity, thus they need more space to continue to produce well.

Consider the textural contrast you want to create in your food garden. Blending not only different foods together but food and flowers is an opportunity to design a visually appetizing garden. Snapdragons are a whimsical, old-fashioned vertical element to mix in with edibles, and the bumblebees flock to them. I love watching the bees earn that pollen, nudging open the sweet snapdragon to reach the prize. They are great flowers to tuck in between lower-growing plants, like compact peppers and between rows of onions, adding random spikes of color to an otherwise uniform area.

Think about the form of a plant, its space and height requirements, and how it will play with its neighbors. Anchor a bed with tall plants in the middle or on the north side of the bed, and let the surrounding plants vary in height.

Find your diagonal views and study them. Consider the space as a whole system, from groundcover to your borrowed views.

If a framework is what supports you in the garden, lean on that. If being freeform suits you, run with it. The best way to create a succession garden is to play with your food (and hopefully flowers and herbs too) making as inviting and productive a space as you possibly can for your local insect population while driving your creativity along the way.

Dos and Don'ts: Interplant, Don't Overplant

Natural ecosystems such as forest edges are great examples of what happens when plants compete for limited resources. In the case of the forest edge, the limiting resource is light. Plants at the edge stretch and bend, reaching out horizontally, elongating their stems vertically, grasping for substantially available light. Plants congregate there because it's their best shot at success: a gap in the forest; the shoreline of a lake; where an expanse of woods meets an open field. Edges are places where light penetrates deeper than in the middle of mature stands of trees.

Garden plants in full sun who are densely interplanted and competing with one another for precious sunlight and nutrients contend with a similar environment. This most often results from an enthusiastic and well-intentioned hand, as well as pressure from uninvited interlopers that germinate in the garden, often labeled weeds. While they eventually become compost, weeds directly

Finding the delicate balance between interplanting and overplanting takes experience. Low-growing flowers like sweet alyssum and fast-maturing vegetables like radish are among the most amenable vegetable garden companions, seen here maturing in between rows of establishing garbanzo beans.

compete for the resources our food needs to thrive. A weedy garden is a great example of an overplanted landscape.

Plant stress occurs when plants are not provided adequate space to acquire the light, water, and nutrients they need, and when stressed, plants are predisposed to disease. Avoid this as you begin to learn interplanting by being clear with your goals for your growing space, and giving your plants the required minimum space they need to do well.

Plants generally become leggy if overplanted, and this can even happen in full sun if they are not spaced properly. Again, overcrowded plants become vulnerable to disease. Powdery mildew happily spreads in densely planted settings like that gorgeous patch of zinnia or the untended squash that's so wild you turned your back on it.

On the flip side, it is possible for plants to be spaced too far apart. Yes, that sounds crazy even to write, but it's true, although rare. I've seen it firsthand. Cauliflower do best 18–24 in. apart, but spacing too far apart can result in a hollow stem, a symptom also observed when the plant is deficient in boron.

It's almost as if the plant needs boundaries, physically touching its neighbors, so as not to stretch itself too far. Now there is a lesson we can take to heart.

Learning how to manipulate your growing space through different types of succession planting will elevate your garden into a more meaningful, more abundant, and more glorious edible playground. The most important thing to remember is to just go for it. Try different flowers and vegetables together, exploring their color and texture and form. Play with interplanting various vegetables, mixing height and maturation rates for a lively and useful interplanted experimental space.

The longer I do this, the more I appreciate the merits of each type of succession planting, applying them variously throughout the garden. The beauty of the garden is that it is a highly personal affair, a direct reflection of the heart and hands of each individual gardener, what their vision of beauty and function look like. And, most importantly, a reflection of what they love to grow and eat.

Flowering early in the season, apple trees provide early beauty and function followed by delicious, crunchy harvests as the weather cools in late summer. Perennial foods offer an annual harvest, planted once.

EDIBLE PERENNIALS IN THE FOOD GARDEN

While it takes more time and resources to establish than a garden of annual flowers and vegetables, true to its name, the perennial food garden offers the promise of fruitful rewards year after year. In this chapter, I will discuss the most commonly cultivated perennial foods found in home gardens, highlighting how they marry together into a robust succession measured in years, including: berries, native shrubs, perennial vegetables, cultivated fruit trees, and native edible trees.

Berries: Early and Mid-Succession Fruits

It's a dream come true: food you plant once that lives on for many years, and is quick to produce, usually in the very first year it's planted. A perennial paradise of berries is easily attainable in the home garden. It provides many added layers and benefits to the landscape, and is at home both in the food garden and mingling with your formal landscape. These plants' flowers provide prized early season pollen and nectar for pollinators in spring. They mature across the growing season, stretching out their harvests from late spring through fall. For all these reasons, berries are one of the most common and beloved fruit groups I know.

    The term "berry" is a bit of a misnomer, because only one of the fruits discussed here is a true berry. For the purposes of this book, however, "berry"

Perennial fruit arrives after berry season, offering mid- and late season successions in the food garden like these plums, early apples, early pears, and elderberries.

comprises what we culturally consume and call berries: strawberries, raspberries, blackberries, and blueberries. Berries of all types are relatively fast to establish and mature—and have an intrinsic proclivity to permanently institutionalize themselves into your landscape with their vigorous and enthusiastic propagation. Plan to find a permanent home for these perennials, a place where their sprawl and delight won't hamper other fruitful endeavors. As you'll see, some berries, like vegetables, cohabitate well, while others work best flying solo.

Strawberries

A relatively short-lived perennial that clones itself enthusiastically by way of runners, the strawberry is a juicy treat that cheerfully makes itself at home practically anywhere in the garden. And while that's a welcome sign, it's also a cautionary tale—a little goes a long way in a very short time with strawberries. I speak from firsthand experience overplanting strawberries, overzealously having ordered 75 bare root plants (25 each of three varieties) for a border in an urban garden. I ended up rehoming more than half the runners the following year because we quickly ran out of space.

June-bearing strawberries are a harbinger of summer's imminent heat. They are the first homegrown berry to reach our eager hands, a temporal hallmark and seasonal reminder that we've made the full trip around the sun once again.

The lowest growing of the common berries, strawberry is actually an aggregate fruit in the rose family, not a true berry at all, but rather a cluster of *achenes*. Each seed you see on the outside of a strawberry is an individual, each little speck a complete fruit, containing a ripened ovary and seed. The red, sweet, mouth-watering goodness that we prize is not the ripe ovary—what we traditionally think of as fruit—but merely an accessory fruit, evolved to draw attention to all those individual fruits so it can successfully proliferate.

Interestingly, strawberries are no longer reproduced by seed. Cultivated strawberries are hybrids, having been cross-bred centuries ago, most notably in England. A larger fruiting variety from Chile and a variety from North America are the parents of modern-day strawberry cultivars, a well-loved and studied horticultural cross. This cross has been bred over and again, for many hundreds of years, for specific traits like fruit size, flavor, maturation time, and in the case of industrial farming, transportability. Strawberries are most commonly sold as bare root plants.

The beauty of strawberries lies in its unassuming form. Dense, low-growing masses of serrated, trifoliate leaflets swiftly creep across a space, practically enveloping it. An ideal border plant anywhere on the landscape, it can be mixed in with cultivated shrubs or perennials, in the partial shade underneath fruit trees, or in its very own dedicated raised bed or children's garden. It expediently establishes into a lush ground cover, and while benefiting from and appreciating some annual maintenance to ensure robust production, even an unmanaged patch will yield plenty of pickings for you and curious wildlife for many years.

CHOOSING STRAWBERRIES FOR YOUR GARDEN

Strawberry Type	Characteristics	Common Varieties	Pros	Cons
June-bearing	Set flower buds in autumn as day length shortens	Cavendish, Earliglow, Honeoye, Jewel	Heavy harvests in short timeframe (two to three weeks)	Extremely vigorous via asexual reproduction through runners
			Excellent flavor	
			Many varieties known to produce large fruits	
			Best type for preserving and canning because of large harvests	
Everbearing	Produce in tandem with June-bearing; flower again in late summer/ early autumn	Beauty, Fort Laramie, Ogallala, Ozark	Longer harvest season than June-bearing	Smaller harvest may make canning challenging
			Lower maintenance, as plants do not produce as many runners	
Day-neutral	Flowering not determined by day length	Tribute, Tristar	Improved everbearing varieties	Similar to everbearing; more appropriate for eating fresh because of reduced harvest sizes in June
			More productive during the cool season	

Strawberry Renovation

Strawberry patches are most productive when maintained through annual renovation. Strawberry renovation is the process of renewing the planting after it's done fruiting. This should be completed within a month of the final June-bearing fruit harvest.

1. Mow down the plants. Be sure not to mow the crown, as that will damage them. We use the same walk-behind gas mower used to mow the paths between our in-ground raised beds.

2. Rake and remove the mulched vegetation from the bed. Within a few weeks, new shoots will emerge from the crowns.

3. After the plants have grown back, thin the bed out. It's important to remove older plants in favor of younger ones. This keeps the bed vigorous and reduces disease pressure.

4. Since a single June-bearing plant can send out over 100 runners (baby plants) annually, thinning out the entire bed is essential.

5. Final plant spacing should be no less than one plant per square foot.

On the continuum of succession gardening, strawberries are the earliest perennials to produce, as fast to produce fruit from bareroot plants as some annual vegetables that take several months to mature and produce. They even produce (albeit modestly) their first year in the garden, which is unusual for a perennial. Clipping first-year blossoms is often recommended, but I cannot recall a new garden where I've done that—and we have always harvested strawberries, even the first year we planted them. It's a true delight, providing nearly instant gratification, and enchanting gardeners young and old.

Raspberries and Blackberries

Raspberry and blackberry closely follow strawberry as early succession perennials, providing delicate fruit that delights the senses, even if just a handful, their first year. These scrumptious aggregate fruits, comprised of clusters of enlarged ovaries assembled together as a mouth-watering treat, benefit from more space than you may think they need. Once established, they send out a plethora of *stolons*, or underground stems, in every direction, seeking to expand their domain. It's best to embrace their gregarious nature, and provide them with a sunny and sprawling permanent home on your landscape, because one thing is certain with these caning fruits: they will sprawl.

Variety awaits the curious gardener with raspberries and blackberries, too. There's summer-fruiting and fall-fruiting, and within that, an array of color and timing to delight those who love to plan. Fall-bearing varieties will produce twice in one year; the first flush of fruit is produced on last year's canes (called *floricanes*), ripening as strawberries sunset and blueberries arrive. The second flush of fruit arrives in autumn on the new growth (*primocanes*). Canes of these shrubs live for two years. In our climate, fall-bearing fruit is often met with early frost, and the harvest is meager relative to summer-bearing canes. In warmer climates, two harvests from fall-bearing plants is both feasible and ideal.

Once floricanes produce a summer crop, they're deadwood, so to speak. Pruning them out is a thorny necessity, but helps maintain a semblance of order. I find raspberry brambles most approachable when I can reach in and access fresh fruit without puncturing a forearm, or entangling any clothing. If my arm returns unscathed, that's a sign the bed has been well-tended.

ABOVE
Strawberries before renovation filling out the ground below our espalier apple trees.

LEFT
Post-renovation, strawberries grow back strong during the second half of summer.

Perhaps the best part is that a little goes a long way in the raspberry bed, and that raspberries and blackberries are as wildly successful at sprawling as strawberries—though rather than obvious, aboveground runners, their new canes erupt surreptitiously, quickly filling out their human-sanctioned living quarters. A passionate and dedicated garden hand is a raspberry patch's friend, with time set aside each year to dig up errant runners in summer, and thin out the patch to a healthy and manageable size when it's dormant.

Worth the Wait: The True (Blue)Berry

The beloved blueberry is irrefutably one of the most striking edible shrubs. The cultivars we gardeners know and love to grow and eat originate from a native shrub endemic to eastern North America, the northern highbush blueberry (*Vaccinum corymbosum*). Blueberries are always in season in one form or another: early spring nectar for native pollinators; petite, waxy foliage and clusters of summer fruit; brilliant fall color and winter interest. Blueberry picking was a

After four years of tending and meager harvests, we began to see more substantial flushes of fruit on our establishing blueberries.

OPPOSITE
Quietly making their way to the next bed, our Double Gold raspberries stealthily spanned a 5-ft. walkway in search of compost-rich soil, erupting a new patch of canes in an attempt to expand their territory into the herb garden.

summer pastime of mine with my mother and sisters, and I don't recall an outing where mild digestive discomfort didn't ensue from noshing my way through the blueberry patch, marginally contributing to the harvest.

The most versatile edible shrub on our landscape has always been the graceful blueberry. Blueberries, like strawberries, are sturdy and confident enough to be food garden accompaniments, as well as outstanding additions to your most formal perennial landscape. With proper soil acidification, blueberries can thrive and produce abundantly for a few decades or more. We quickly replaced our overgrown cedar shrubs in the first home we ever owned with blueberries. Interplanted with strawberries, they were the foundation of our edible urban front yard garden.

While it seems like blueberry would be quick to mature, this slow-growing shrub is a later-succession fruit in the food garden. We've observed over 15 years of growing them that they require the most patience of all the berries we grow—even more than some of our fruit trees. It is hard to believe such a small shrub can take so long to mature, but it can take anywhere from five to ten years for blueberries to come into full production. So, it's best to temper expectations, and plant them now if you can.

Blueberries remain a prized commodity in our home garden, and I remain not-so-patiently waiting for those mythically overflowing bowls of blueberries, the five to ten pounds each of our twelve shrubs will someday produce annually. Despite all the soil amendments, space, and four-plus years, we've only picked handfuls here and there to date. I'm sure we haven't been completely honest with each other when we've snitched a few ripe ones here and there, children and parents alike. It does seem the tide has finally turned in year four, with growing harvests, so perhaps the furtive berry sneaking will lessen for all of the blueberry lovers in our home. (That's four of four family members, so competition for the fruit remains stiff.)

If you want blueberries and can't establish all your perennial fruits in one season, I highly recommend prioritizing them ahead of faster maturing fruit like strawberries and raspberries; give them that extra time to root down and produce for you.

Blueberries are a great way to stagger the harvest, adding more layers of succession to your landscape, as their fruit can mature over a range of weeks. For successful cross-pollination, you need at least two plants, so selecting an early, mid-, and late season variety for your blueberry planting will naturally stretch your harvest as long as possible. While I think of blueberries as a gift for the northern winters we survive, there are plenty of varieties suited to warmer southern winters too. There's a blueberry for everyone.

PERENNIAL FRUIT SUCCESSION

						🍐	🍐🍐	🍐🍐🍐
				🍑	🍑🍑	🍑🍑🍑	🍑🍑🍑	🍑🍑🍑
			🍑	🍑🍑	🍑🍑🍑	🍑🍑🍑🍑🍑	🍑🍑🍑🍑🍑	🍑🍑🍑🍑🍑
			🍎	🍎🍎	🍎🍎🍎🍎	🍎🍎🍎🍎	🍎🍎🍎🍎	🍎🍎🍎🍎
		🫐🫐	🫐🫐🫐	🫐🫐🫐🫐	🫐🫐🫐🫐	🫐🫐🫐🫐	🫐🫐🫐🫐	🫐🫐🫐🫐
	🍇	🍇🍇	🍇🍇🍇	🍇🍇🍇🍇	🍇🍇🍇🍇	🍇🍇🍇🍇🍇	🍇🍇🍇🍇🍇	🍇🍇🍇🍇🍇
🍓	🍓🍓	🍓🍓🍓	🍓🍓🍓🍓	🍓🍓🍓🍓	🍓🍓🍓🍓	🍓🍓🍓🍓	🍓🍓🍓🍓	🍓🍓🍓🍓
YEAR 1	**YEAR 2**	**YEAR 3**	**YEAR 4**	**YEAR 5**	**YEAR 6**	**YEAR 7**	**YEAR 8**	**YEAR 9**

From top to bottom: pears, apricots, plums, apples, blueberries, raspberries, strawberries.

The Early Season Honeyberry

Honeyberry is a novel true berry gaining in popularity. In cold climates, where the wait for life to return in spring is long, honeyberries begin maturing around or before the earliest strawberries, a concept that seems too good to be true. Honeyberries are a relative of honeysuckle, though to date, they do not demonstrate that plant's aggressive or invasive qualities. Breeding of Japanese and Russian varieties has been ongoing for the past 20 years or so, with the goal of improving flavor. You can source this unique berry easily through many mail-order catalogs. (See appendix for a list of sources.)

Honeyberries, like blueberries and fruit trees, need more than one plant for successful pollination—and to produce a good harvest. Be sure to research which varieties might do well for you. As with many other fruits, you can stretch your harvest with early, mid-, and late season varieties. While they take a few years to produce, both their flowers and fruit arrive very early, even by northern garden standards. If your season is short, this may fill a local fruit gap during your hungry season in late spring.

Honeyberry flowers in clusters of pairs, and fruits mature from mid-June through early July in Minnesota.

Native Edible Shrubs

Branching out beyond cultivated fruit are native shrubs. Mother Nature has perfected these plants over the eons, and they continue to delight humans and wildlife alike. Well adapted to their endemic ranges, native edible shrubs bring high ecological value to any landscape, providing food and forage for insects and wildlife that have evolved to flourish right alongside these plants.

My top native edible shrub picks for the home garden are all true natives, not "nativars," or cultivars of natives. Order American elderberry (*Sambucus canadensis*) if you live in the eastern half of North America, and resist those fancy cultivars marketing unique lace or variegated foliage and boosting larger berries. They will all produce berries for you, but when you add true native edibles, you contribute significantly to your local ecosystem. These nutritional powerhouses provide familiar flavors and potency that haven't been manipulated or degraded by breeding for specific traits. They also nourish your native insect population, who will nourish you in return with robustly pollinated fruit.

Foraging native hazelnuts in an open oak woodland in early September. The largest seeds will be saved, stratified, and germinated to add to our home landscape.

NATIVE SHRUBS FOR YOUR SUCCESSION GARDEN

Shrub	Native Range	Years to Fruit	Height	Light Requirements	Cultural Considerations
Gooseberry (*Ribes* spp.)	Northern United States and Canada	3–5 years	6–12 ft.	Part shade to full sun	Alternate host of white pine blister rust, which kills pine trees; forage favored by birds and flowers by native pollinators
Saskatoon berry or juneberry (*Amelanchier alnifolia*)	North America	3–5 years	4–15 ft.	Full sun	Early flowers and fruit highly nutritious and milder than blueberry; underutilized
Chokeberry (*Aronia* spp.)	Eastern half of North America	5 years	5–6 ft.	Full sun to part shade	More astringent than elderberry, but higher in antioxidants
Elderberry (*Sambucus* spp.)	Species vary across North America	3–5 years	5–12 ft., depending on species	Full sun to part shade	Excellent edible and wildlife plant; can be aggressive, so best suited to sites with room to spread
American hazelnut (*Corylus americana*)	Eastern half of North America	4–7 years	6–10 ft.	Full sun to part shade	Smaller than commercialized European hybrids, but an excellent addition as an understory shrub in open woodlands or forest edges
Lingonberry (*Vaccinium vitis-idaea* var. *minus*)	Northern United States and Canada	4–7 years	12–18 in.	Full sun to part shade	Low-growing alternative to cranberry; thrives in wet, acidic soil; relative of blueberry and cranberry
American cranberrybush or highbush cranberry (*Viburnum opulus* var. *americanum*)	Northern contiguous United States and Canada	5 years	6–10 ft.	Full sun to part shade	Excellent wildlife shrub; fruit tart but edible

Edible Oddities: Perennial Vegetables

Perennial vegetables typically kick off the growing season very early, and in tasty fashion. From rhubarb to sorrel, asparagus to chives, perennial vegetables span the breadth of climates, making friends among the coldest and most challenging gardeners. These reliable vegetables poke their edible stems up from the still-chilled earth as the first hint of spring slowly warms the bare land, enticing the top few inches of soil to awaken before deeply rooted trees, shrubs, and prairie plants emerge from dormancy.

It might surprise you to consider some of the cornerstones of our diets are, in fact, perennials in their native climates. The tomato, pepper, eggplant, and potato, all close cousins, hail from the nightshade family. Kin to several other

Common elderberry provides more than just food—it lures cedar waxwings to our property in late summer. With plenty of berries to share, native edibles provide the best possible nutrition for wildlife.

plants that are poisonous, their own foliage is mildly toxic to humans. We grow these vegetables largely as annuals, even in climates where they persist for a few years or more. The real stars of the perennial vegetable garden, however, are plants we enjoy as tried and true perennials, ones that have mostly remained seasonal delicacies over the past few centuries. They are foods eaten best in spring, when fresh and close to the land.

Perennials are such a gift of time, for the longer you have them in the ground, the deeper their roots grow, and the more they produce. For perennial foods to nourish us year after year, we must allow them to nourish themselves, replenishing their sugar reserves in their root systems so they can reliably spring forth the following year, showering us with another harvest. So, as you manage your even-aged asparagus patch, clear cutting the emerging spears day after day for several weeks in spring, there comes a moment when you say to the plant, "Enough, thank you. Now it's time to feed yourself for the rest of the growing season."

Rhubarb stems are ground-level garden art—and a harbinger of the commencing growing season.

Edible Shoots: Asparagus, Rhubarb, and More

With roots penetrating downwards of 10 ft. in well-drained soil, asparagus sends up spear after spear directly out of the soil, visibly growing inches per day. Left unattended, the first spears will continue to grow up and begin their transformation into a massive hedge of fern-like branches. The savvy gardener knows to take their harvesting knife to the asparagus patch at least once a day to pluck the most recent eruptions from deep below the earth. Harbingers of Mother's Day and fruit blossoms, this timekeeper heralds the first major vegetable harvest of the outdoor growing season. Where established, asparagus is often the first food harvested in spring, launching the growing season into life with perennial ease.

Another vegetable stem we love (or have learned to love, by way of sugar and strawberries) is rhubarb, probably your parents' or grandparents' least favorite vegetable. Rhubarb is a hardy perennial that even the deer don't sample, regardless of how unappealing the current landscape buffet may be. If that's not an indication as to why serving it as a vegetable (once customary) turned so many people off, I don't know what is. But it's no wonder it's a staple of old-time gardens, common on old homesteads, and valued by so many cold climate gardeners. It grows well with very little pest or disease pressure, and simply comes back stronger year after year.

Rhubarb is similarly useful to asparagus with its early season, crinkled tufts of fiery foliage that signal the end of snow is nigh. This perennial vegetable rolls in quite a bit before strawberries even finish setting fruit, and long before its culinary companion is ripe. Luckily, you can harvest rhubarb over a long period, and it freezes well, so it's still very much in season for strawberry-rhubarb everything. It deserves a permanent location in every home landscape, and if you're anything like us, you'll love knowing it's there and waiting, low maintenance, deer-resistant, and pest-free.

Fascinating vegetables grow in many climates—things like tree collards, which persist for several years as perennial shrubs, similar in taste to kale, their botanical relative. Or sea kale, another fascinating brassica endemic to the North Atlantic coastline. Other marvelously curious and persistent foods grow in milder climates than ours. Again, the theme is edible leafy parts. It's what seems to bind these perennial edibles together and make them utilitarian additions to home gardens. And remember, it's best to make use of the earliest growth, and save plenty for the plant to store up energy to persist for years.

The Late Succession Garden: Fruit Trees

The old-growth component of the food garden is fruit and nut trees. A late succession food source, fruit trees take time to establish roots deep enough to sustain themselves before blessing your home with abundance. At least a modest handful of years (often more) are behind those first fruits you ceremoniously pluck from the tree and share with your loved ones. To call them prized would be a gross understatement— seasons of yearning are encapsulated in that first bite. It starts with observation and cheering on, rooting, and tending; proceeds to blossoming; and finally ripening, and delicate devouring.

Maybe it's just me, but there's something deeply sentimental about fruit trees on the landscape. Maybe it's the abandoned homestead apple tree that

Pears dripping from mature trees in early autumn are late succession fruit, taking between 5–10 years to bear.

catches my attention in full spring blossom and then again in fall, dripping with unnamed varieties that go unpruned, and sometimes unnoticed—but especially untamed in all the best ways. I feel a deep reverence in the presence of such trees. It might be because I didn't grow up with a fruit tree in my yard, or even in our neighborhood that I can recall. There was precious little wild fruit to treasure other than wild raspberries (so cherished by my sister, neighbor, and me that we took turns pretending to be raspberry royalty while the others gathered the harvest). I am in awe of how fruit trees endure, their longevity and continuity, and their awakening each spring.

Fruit trees are the essence of abundance, and there is a fruit tree for every garden and every climate. From apple and pear to peach, apricot, nectarine, plum, cherry, and—if your climate is gloriously balmy—lemon, grapefruit, mandarin, and avocado, just to name a few. While yours might be a fig and mine an apple, the abundant diversity of root stock and chilling hour requirements offers limitless opportunities. In containers or in the ground, any and every nook and cranny makes a good space for a fruit tree, in my opinion.

Planting a fruit orchard is also one of the great cornerstones of the homestead philosophy, and a welcome addition to any size landscape, urban or rural. With their wide range of chilling requirements, apples, for example, can be grown just about anywhere, from zone 2 all the way up to zone 9, though not all apples grow well in every location. We can grow Honeycrisp, Zestar, and Haralson, all delightful, locally-bred varieties that require cold winters to thrive. These apples don't reciprocate in warmer climates, and if that's where you live, seek out those with shorter chill hours, such as Anna, Pink Lady, Fuji, and Granny Smith.

Playing with fruit trees in your garden brings limitless joy and potential. Locate your warmest microclimate, likely against the southeast side of your house, and push your zone, tucking in a tender specimen as an exercise in curiosity. For us, this would be a peach, nectarine, or cold-hardy fig. You could also invest in a large, sturdy pot, and enjoy dwarf citrus even in a northern climate, sharing space with it in fall and winter as a houseplant, and using it as a patio statement in spring and summer.

OPPOSITE
One apple tree can produce over 150 lbs. of fruit at maturity. Establish them early to maximize your fruit successions in your food garden.

Rooting for Perennials

As with annual vegetables, where cultivars exist for perennial foods, so do choices. Do you want to extend your strawberry season by planting early, mid-, and late June-bearing varieties? Maybe you'd like to try ever-bearing strawberries that dribble out the harvests for an even longer stretch? The first step to achieving succession in your perennial food garden is to understand the space needs of each of the foods you'd like to grow so you can match those dreams with the reality of your space and light constraints. Providing appropriate resources is a necessary practice for successful and realistic garden planning, which I'll discuss more later in this chapter.

Once you understand your site's potential, you can determine the number of trees or shrubs you can add. That's when the fun really begins, paging through catalogs and websites, dreaming and scheming of how to extend the fruit season even longer in your own backyard through careful orchard curation.

Fruit trees can be cultivated in varying sizes, determined by the root stock you choose at the time of purchase. While the best use of any space is espalier dwarf trees, semi-dwarf trees are a fabulous option, as they do not require any staking when open grown. Dwarf root stock, the roots on which the variety is cloned, requires staking for apple trees because the root system can't support the weight of a full harvest. It has the benefit, however, of being easy to harvest because it tops out around 8 ft. Long gone are the 20 ft. tall apple trees and awkward ladders and buckets; simply invest in dwarf trees and your fruit will be very much within reach.

While most perennial edibles are sold bare-root (and that's your best value as a consumer), a few arrive in large bundles. As such, it can become a logistical challenge to purchase a wide variety when the minimum order is, say, 25 bareroot plants. This is the case with strawberries, which, if spaced correctly, will cover at least 100 sq. ft. In an urban setting, this quantity might best be split with a friend, neighbor, or interplanted in and around existing shrubs in sunny spots throughout your yard.

The Old Growth of the Food Garden: Native Edible Trees

Out beyond the decade it takes for your apples, pears, cherries, plums, and other cultivated fruit trees mature, there is yet another layer of opportunity to incorporate food into your landscape, though this timeline is longer still. Along with the patience for harvest comes a legacy that lasts many decades, a living mark of your passion for and commitment to your landscape.

From walnut to pecan, wild plum to mulberry, there are many nutritious and delicious foods harvested from trees that grow the world over, and every location has a plethora of well-loved endemic fruits and nuts. Native birds and mammals small and large already know the best places to forage these nutritious feasts, in tune with the ebb and flow of the seasons.

Native trees are the final element in deepening the perennial succession garden, bringing nourishment in such an abundance of forms that it might feel like a fairy tale. Young woody plants, like those adorable little whips you can order bareroot from places like Arbor Day Foundation, will take a decade or more just to reach your stature, let alone full maturity. These are the furthest thing from the quick garden radish, but they are the balance to that immediacy, plants that embrace the slowdown, inviting us to bear witness to the tremendous effort trees make to reach maturity and set fruit.

The beauty of foraging for wild edibles to plant is that you get to enjoy the fruits first. We foraged local American plums in early September, seen here (top right) along with an average September harvest. We savored their sweetness, and saved the biggest seeds to germinate the following winter.

NATIVE EDIBLE FRUIT TREES FOR SUCCESSION GARDENS

Tree	Native Range	Years to Maturity (Bearing Fruit) / Height	Light Requirements	Cultural Considerations
American plum (*Prunus americana*)	East coast of North America to interior west	10–12 years / 12 ft.	Part sun to full sun	Can form thickets through root suckers; delicate and tasty fruit
Serviceberry (*Amelanchier* spp.)	Eastern half of United States, including plains states and Texas	5–10 years / 15–25 ft.	Full sun to part shade	Single or multi-stemmed, open growth habit; edible fruits best left for the birds
Red mulberry (*Morus rubra*)	Eastern half of United States, excluding far northern states	10 years / 35–50 ft.	Full sun to part shade	Dioecious, needing male and female trees to fruit; can drop an excess of seeds, resulting in heavy (weed) seed bed; edible and enjoyed by birds and humans
Common persimmon (*Diospyros virginiana*)	Eastern United States to zone 5, excluding northern New England	10 or more years / 15–20 ft., in excess of 50 ft. in rich soils	Part shade	Dioecious, needing male and female trees to fruit; culinarily versatile, used for centuries by Indigenous cultures
Paw paw (*Asimina triloba*)	Eastern United States to zone 5, excluding northern New England	4–8 years to bear fruit / 15–30 ft.	Full sun to part shade	Understory tree with delicate fruit; considered "the mango of the temperate forest"

While the pollination of perennial fruit in our garden happens within a span of about a month in spring, its harvests spread out across early summer through autumn. A slow trickle of sweetness commences along with June's midsummer strawberries and flows right into July's berries, with August's plums, apricots, elderberries, and early apples and hazelnuts following right behind. The leaf-raking months are the final late season fruit succession, with late season apples and pears and the errant handful of fall raspberries gracing our harvest baskets. The combination of these perennial edibles creates a long-term edible succession garden.

This glorious orchestra of fruit anchors the food garden, carrying us through not only the four seasons, but across the years and decades. I can't think of a more beautifully functional component of the landscape than edible perennials, formalized or naturalized. They tether us to seasonal eating and, as always, help spread the abundance for longer, and thus offer us some breathing room as we work to enjoy and preserve that which our land provides.

Establishing Perennial Edibles

Perennial food is integral to exponentially robust offerings from a garden of any size. It wasn't until we bought our first home that perennial foods became a possibility for my family. We focused on fruit production, and the joy of planning and planting fruit trees was a highlight of re-imagining our postage-stamp urban landscape. And as we moved into new homes, fruit trees were always the first food to be planted, as bareroot trees in spring. Because they take longer to produce, it is imperative that priority be given to their establishment. In this section, I'll discuss what to consider when adding perennial edibles to your landscape.

Trees First

Consider placement of fruit trees carefully. They are, in essence, the overstory of your food garden, especially if they will be open grown and in the same vicinity where you plan to grow your vegetables. Their summer shadows will dance around, reducing sunlight—and thus potential food production—in the neighboring understory. In the early years, light under a fruit tree will be ample for

interplanting, but as the seasons pass, available sunlight will diminish. While space might not be at a premium in our sprawling suburban garden, we planned our fruit orchard carefully to keep its shadows at the edges.

Because deer pressure is high in our area, all of our prized fruit trees are planted within the safety and protection of our rabbit- and deer-proof fence. Thirteen fruit trees reside inside our ⅛-acre garden oasis. If open grown, those trees would occupy nearly ⅓ of our available growing space inside the deer fence.

We chose to espalier our fruit trees, lining the northern and eastern perimeters of the food garden with their trained branches. Because of this training, the impact of their shadows on the rest of the garden beds is negligible. At maturity, they will produce fruit and provide a screen, enclosing our garden in an edible hedge.

Garden Art: Espalier Fruit Trees

More seeds for edible landscaping around our property, we high graded the largest foraged hazelnuts for germinating in the winter, and roasted and enjoyed the rest.

Espalier is a method of training plants that hails from Europe. Dating back several centuries, it is similar to the method in which grapes are grown. In the simplest terms, you train a plant to grow in two dimensions, laterally and vertically. By doing this, you restrict growth to only the horizontal and vertical planes, while minimizing the third dimension. These higher density plantings encourage robust harvests, but occupy a minimum of garden space. The space you'll gain is ample for other foods to flourish. It is the best of both worlds, where fruit production blends with functional beauty, maximizing a growing space.

Deer- and rabbit-proof fence, seen here behind our espaliered apricot, protects against two of the most voracious herbivores native to our area, and proved an invaluable investment before we sunk the first spade into our prized food garden.

Dwarf apple trees planted densely create a hedge and produce plentifully in a small space, a more informal but equally useful training method to espalier.

It's a stellar approach for an urban setting, and for us, it's also one of the many ways our family's heritage shines in our garden, with several generations before us having joyfully maximized urban growing spaces through espalier. Our first fruit trees (three apples) lined the south side of our city home, along the foundation. We were blessed with a one-story home to our south, which meant sunny summer days that would encourage these trees to eventually produce our very first homegrown fruits. Espalier fruit trees generally take longer to fruit, but the space saved in our urban lot allowed us to grow strawberries, herbs, watermelons, peppers, and eggplants in front of the establishing trees—plus an entire front yard dedicated to food and flowers, which otherwise would have been solely a fruit orchard.

ESPALIER EXAMPLES

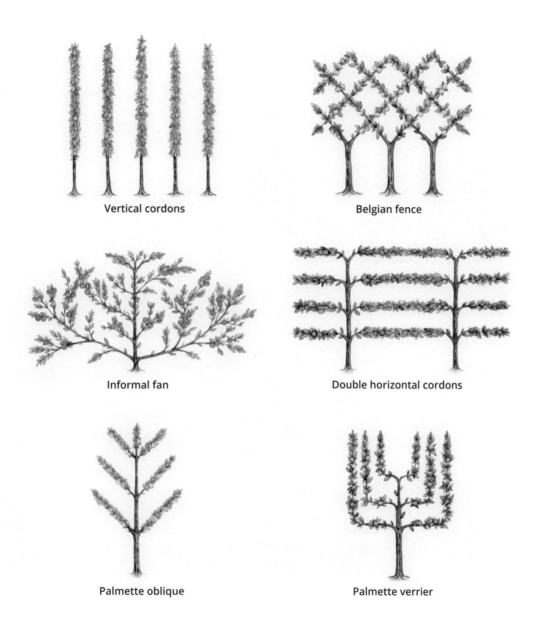

Vertical cordons

Belgian fence

Informal fan

Double horizontal cordons

Palmette oblique

Palmette verrier

Functional, edible art, espalier fruit trees save space while adding undeniable beauty and unique form to the food garden.

Tips for Buying Perennial Edibles

Many fruit trees require cross-pollination for good productivity. The most common fruit trees, such as apples, pears, and plums, need more than one tree for robust pollination. Nectarines, apricots, and peaches are self-fertile, and though they benefit from cross-pollination, they don't require it. Cross-pollination will increase overall yield. Many old homesteads strategically added crabapples to their property, as they are excellent for this purpose.

Catalogs also list good cross pollinator options. Research and plan for overlapping bloom periods, so pollen will reliably be shared between and among your trees by the pollinators. If you only have room for one tree, be sure to order a self-fertile variety. If you're an urban gardener, there's a very strong chance a neighbor has a pollen and nectar source that will work with your apple tree too.

RIGHT
Despite a 3 ft. path, this stretch of our perennial garden grows into a lush tunnel by late August, the asparagus naturally leaning into the spacious edges we provide it. We wouldn't want it any other way.

OPPOSITE
Our north orchard is a row of espalier apricot, plum, and apple trees. Plums and apples were the first to bear fruit in year four, and apricots won't be long now.

Shrubs and Berries

Edible shrubs and other perennials are the next layer of succession to plan for and plant as you establish your garden. We nestled these early and mid-succession perennials directly in front of our espalier orchard, lining the northern 15 ft. with edible perennials. Added among the fruit trees were raspberries, blueberries, strawberries, and asparagus. While not as tall as the trees, these plantings were set between slightly wider paths than our vegetable beds to accommodate their mature stature and late summer sprawl. Their limbs dependably flop and stretch out beyond their beds in late summer for a little breathing room and to capture that extra sunlight.

Growing Asparagus from Seed

Asparagus is a fantastic vegetable to extend your garden season, and it's most commonly available as one-year crowns. Like other perennials, its root system must establish before harvesting can begin. For one-year crowns, it's common to wait one to three years after planting before beginning to harvest this early spring delicacy.

True to my philosophy, the trick is to get an extra early start by sowing from seed. Growing asparagus from seed lessens that wait by a year or more. The more growing time you provide a seedling its first year, the larger its root system, and the stronger it will be as a transplant. By sowing from seed early (late January), we were able to begin meager harvests at the one-year crown stage. By year two, we were harvesting for three weeks straight.

Sow asparagus seeds in soil blocks or plug trays twelve to fourteen weeks before your last spring frost. Germination may take up to three weeks, so I recommend closer to fourteen weeks for cold climates. Pot them up as needed, likely around four to six weeks after germination. We potted into a 3-in. pot, and transplanted those into the garden after danger of frost had passed in early May. To transplant, in well-amended soil, use a post-hole digger to make a hole at least 12 in. deep. Space transplants 18 in. apart in rows of 3–4 ft. Add slow-release organic fertilizer and plant. As the seedling grows, continue to fill in the hole with soil. Rain also helps with this, with no effort on your part. Broadcast fertilize every autumn, and top dress with compost.

Hazelnuts tolerate some shade, so we happily added a handful of seedlings to our front yard landscape, hoping to enjoy our own nut harvest many years down the road.

FAR RIGHT Blueberry production meagerly increasing with each passing year as the plants inch toward maturity. In this particular year, the raspberries dwarfed the blueberries.

With so many options among berries, fruit trees, and native edible shrubs, how do you narrow down your choices when planning your garden? I wish I could tell you this process is straightforward—it is anything but. New varieties come to market annually, enticing us with their glossy photos and well-crafted blurbs, mentally nudging us to find creative ways to stretch our dedicated growing space further. Being clear about what you love to eat and what brings you the most joy is the first and most important step in selecting what to add.

As you bring perennial foods into your garden, you are committing to a longer-term relationship with your food. The more diversity you plan for, the more resilient you and the garden will become against the inevitable snags that will trip you up in the short term. A fruit tree may die, and you will be left with an opportunity to try a different variety. Pest pressure may be challenging for certain plants. Through the years, you will find what varieties work best for your own microclimate. But add perennial foods as soon as your space and budget allow, and take joy in providing your establishing plants with ideal growing conditions as best you can. Think of it as an investment in the seasons within seasons of food that await your careful cultivation.

Cross-sections of the garden reveal the layers of the seasons melding together. Asparagus in the rear anchors this view with its playful fronds, while onions, brassicas, peppers, and flowers build a lush view in the foreground.

VEGETABLE SUCCESSIONS IN THE FOOD GARDEN

The combination of beauty and function is the cornerstone of the succession food garden. The most practical garden that I can think of nourishes us both literally and figuratively. The most productive succession garden stitches together these elements in a near-constant arrival and departure of food and flowers from the earliest growing days to well beyond the first fall frost, the end of the traditional growing season. Management of this garden balances indoor seed starting with direct sowing; it leans on stellar recordkeeping and scouring catalogs to source as much variability as possible to keep the harvests abundant and diverse throughout the growing season.

The layers of the food garden must be reflective of your needs as the gardener, the garden filled with your favorite foods and most beloved flowers. In order to maintain such a productive garden, you must maintain a very active hand all season long. Varieties that keep you curious and motivated to continually sow come summer's brutal heat, when the garden already feels fully planted, are imperative to bountiful succession planting and extended harvests.

By "all season long," I mean nearly every month of the year, even in the cold and short midwestern growing season. Despite averaging fewer than 5 months in our growing season, we spend more than 10 months of the year working on the garden in one form or another, from seed sourcing and planning in the off-season; to working with low tunnels during late winter; to preparing and storing vegetables in late fall in our root cellars. I partition our annual vegetables

Growing food produces a gorgeous tapestry of edible landscaping that ebbs and flows with the changing seasons and the gardener's palate.

into three main types of plantings: quick succession, midseason succession (also known as middle or main succession), and late season succession. These three categories reflect the length of time it takes a vegetable to mature, producing the desirable food to harvest.

Within these types, the timing and planting across the growing season may be somewhat fluid—especially midseason successions, whose window is broadest. Understanding how these vegetables suit different seasons of the year, how much time they need to mature, and when to plant them will provide you a springboard for creating the edible palate of your dreams.

While some vegetables wake up and rush to the finish line, others are finicky and require just the right soil temperatures and nutrients to thrive. Still others command your attention, time, and space for many months before maturing. Understanding how the various layers of a food garden work harmoniously to produce a cornucopia in each season, and across the growing season, will take your food garden to the next level.

TYPES OF SUCCESSIONS

Succession Type	Time to Maturity	Crop
Quick succession	Under 2 months	Arugula, cilantro, kale, kohlrabi, leafy greens, mustard greens, radishes, spinach
Midseason succession	2–3 months	Corn, cucumber, eggplant, green beans, head lettuce, peppers, summer squash, tomatoes
Late season succession	3–5 months	Brussels sprouts, cabbage, garlic, leeks, onions, potatoes, winter squash
Generalist	2–3 months	Beets, broccoli, cabbage, carrots, herbs, kale

Quick Successions: Garden Fast Food

Remarkably, there is a robust group of vegetables that completes its entire life cycle in mere weeks. The concept of fast food coming straight from the earth is a blessing, and if speed weren't enough of a draw, these vegetables are also prized because they thrive in the cool season. It's during those weeks in spring and again in autumn, when you feel like gardening can't start or is already over, that these plants come waltzing in, dazzling you with their expedient growth and powerful flavors.

The Garden Radish

First to our table almost every spring is the garden radish. It is reliable in cold soils, faster in slightly warmer conditions, and with ample moisture and nutrients, it forms an edible root in mere weeks. Many varieties mature in a month. The entire plant is edible: the root is most commonly prepared, and can be enjoyed raw, fermented, or roasted; the leaves can be lightly steamed or sautéed to go along with any number of dishes, and the seed pod is its own culinary adventure. In early spring, it feels seasonally appropriate to ceremoniously prepare and honor the entire plant, savoring its pungent flavors produced while spring frosts linger, and well before most anything else graces our plates.

Radishes aren't just the round, symmetrical orbs you see in the produce section. While the littlest ones produce the fastest, radishes come in many colors and shapes, and seed catalogs open a wide world of possibilities. I love that radishes don't require much space or time to study their fitness in our gardens and on our plates, and we continue to unearth new favorites every year.

Mighty Greens

Brassicas, a large group of some of the most nutritiously dense, alkaline vegetables, include cabbage, kale, broccoli, cauliflower, kohlrabi, and Brussels sprouts, all variations of the same species, *Brassica oleracea*. Thousands of years of cultivation have yielded specific traits such as unique leafy production (kale), immature inflorescence (broccoli), and an engorged, bulbing stem (kohlrabi).

Radishes, from left to right: Viola, French Breakfast, Shunkyo Semi-Long, White Icicle, KN-Bravo, watermelon (Red Meat or Beauty Heart), and Black Spanish.

Many of these leafy plants make an excellent quick succession in both the early and late season. Because we harvest quick succession brassicas in their vegetative state and not when flowering, they naturally mature faster. Mid- and late season successions, on the other hand, must first accrue sufficient vegetative mass to flower, set fruit, and ripen.

These quick succession leafy crops are also frost tolerant, making them suitable to a wide variety of conditions. Many a northern climate gardener is no stranger to weather whiplash in spring. From 75°F (24°C) and summery one day to dustings of snow the next, these crops shake off meteorological fluctuations effortlessly, thanks in large part to their quick maturity. And frost tolerant crops are well-suited to the early season because they can be sown or transplanted in the ground well before the last spring frost, with protection.

PLANTING TIPS FOR QUICK GREENS

Brassica	Planting Tips
Arugula*	Peppery, pungent green easy to grow from seed; direct sow as soon as soil can be worked, or for earliest harvests, sow indoors and transplant in early spring under cover
Spinach*	Speedy green well-suited to early spring and late autumn; direct sow right into the garden; will yield faster transplanted from starts as early as possible; does not thrive in hot soils
Mustard greens	Another fast peppery green, most often cooked; very heat sensitive, so plant early in spring and again in fall; indoor sowing will propel spring offerings; direct seeding in late summer also suitable
Bok choy	Type of Chinese cabbage developed as loose head; crunchy stems and deep green foliage with mild flavor and two distinct textures; most heat tolerant of quick successions, but sow indoors and transplant for earliest harvests
Leaf lettuce*	Another easy food that grows well direct sown; one month to baby greens, two months to larger leaf lettuce; moderately tolerant of warm weather, but for longest season, start as early as possible in spring
Kale	Must-explore brassica, not just because it's finally mainstream; act as garden art, arrive extra early, persist all season, and grow sweeter after fall frost; among our earliest and latest crops, all harvested from the same plant

Interplant between rows of young, slower growing brassicas such as cabbage, broccoli, or cauliflower.

What continues to surprise me is how well these leafy greens grow during those mentally battering cold weeks of late winter. I have marveled as each of these frost tolerant greens has survived hard spring frosts of 25°F (−4°C) uncovered. The best way to fine tune your gardening skills and your hardiest garden companions is to be willing to expose plants to weather extremes. If you're willing to lose a few from time to time, the knowledge and confidence gained from personal experience outweighs all those short-term losses.

Kohlrabi: The Fastest Brassica

A perfect spring interplanting, spinach is direct sown between young cabbages, where it thrives and delights for weeks on end, fizzling out before the cabbages close the canopy.

My very favorite early season brassica is kohlrabi. It's a uniquely shaped, juicy yet crunchy, fun to grow vegetable that is ready to harvest in under two months when transplanted as four-week-old seedlings. It is the largest and crunchiest early season vegetable I know, and it is sure to start a conversation with an uninitiated houseguest or friend. Like radishes and all brassicas, the leaves are edible too, a boon to early season harvests where greens might otherwise be in short supply. Kohlrabi leaves can easily stand in for kale. The entire plant will fill our plates fresh from the garden before the last spring frost.

Over several years of trial, I have found hybrid varieties perform more consistently and in quicker succession than open-pollinated varieties. Anecdotally, the hybrid varieties seem to focus on bulbing earlier, and with less emphasis on vegetative growth compared to their open-pollinated relatives. In our household, we call this hybrid vigor.

Planting these early maturing vegetables in quick succession, and sowing them at least once a month (as space allows) from late winter through early spring provides a steady stream of beautiful and delicious conversation starters until summer's first blast of heat signals their imminent departure. Hopefully, by then, they are close to maturity, and you've got a plan for what will replace them in the coming weeks. You can and should repeat this entire process in the middle of summer, about eight to ten weeks ahead of first fall frost. That constitutes a well-timed fall garden, extending the growing season beyond beloved frost-intolerant favorites like tomatoes, beans, squash, and potatoes.

The theme in these quick successions is timing, sowing the seeds early and often enough so the plants thrive when most other commonly grown vegetables would simply scoff at an invitation into the garden. Every year brings unique climate challenges, so planting a variety of early season crops is insurance that you will harvest something grown well, regardless of environmental conditions.

ABOVE
Kohlrabi is a speedy and fun spring crop that is delicious raw or cooked, and eagerly takes on the flavor of dressings.

OPPOSITE
By early- to mid-June, our harvests include several types of brassicas we've planted indoors starting in late February.

Early and Late Successions: A Brassica for Every Season

Brassicas are as diverse in their shape, texture, and flavor as in their days to maturity. While varying wildly in form, this group of vegetables encompasses some must-grow early spring food crops. Here in our northern garden, we have brassicas that grow or persist in one form or another from late March until early December, weathering the seasons quite unlike any other vegetable we know of. Thanks to this, their many varieties are the foundation of our succession food garden's prosperity. Our list of favorites continues to grow, while we reserve space to explore new-to-us varieties in the quest for that early, extended, steady, and diverse stream of food production in our growing climate.

All these forms of *Brassica oleracea* have been cultivated and bred for various plant parts, eventually becoming many of the mainstream vegetables we rely on daily.

BRASSICAS FOR EARLY AND LATE SEASON SUCCESSIONS

Brassica	Planting Tips
Arugula	Purported all-season star, but bolts quickly for me in summer; can be added when you remember if you lack the stamina to sow all season
Kohlrabi	Quick, unique vegetable; texture and size impress in a matter of weeks; we grow a few times in spring and in fall since it matures in well under three months, and faster in ideal conditions
Broccoli	Should be specified as early or late season on seed packet or in description; some varieties happier in shoulder season, while others widely adapted; earliest can be sown eight to ten weeks before last frost, and set out under row cover six weeks before that date, as soon as soils are warmer than 50°F (10°C); we grow broccoli throughout our entire growing season
Broccoli raab	Speedy early and late season addition, ready in same time frame as kohlrabi; sow six to eight weeks before last and first frost
Radishes	Direct sow as early as soil can be worked, with soil temperatures above 40°F (4°C); sow weekly for next four to six weeks, until mid- to late spring; repeat in late summer, starting six to eight weeks before first frost, sowing weekly until two weeks to first frost, to extend harvest into fall
Cabbage	Some cabbage varieties can be used as super early spring crop with kohlrabi and broccoli, and again in fall; days to maturity vary widely with cabbages; red cabbage generally slower to mature than green; read descriptions closely before purchasing or sowing; like broccoli, we grow cabbage throughout the growing season here
Cauliflower	Whether white, green, or purple, cauliflower generally does not fare well in summer heat; best cherished as slow but rewarding crop, starting early for either spring or fall garden; finicky, but don't give up (our success rate is about 75 percent with transplants); for fall, sow twelve to fourteen weeks before first fall frost and transplant four weeks later
Romanesco	Meatiest of broccoli varieties we grow; true shoulder season variety, best planted as early as possible; for fall, indoor sow at least ten to twelve weeks before first fall frost
Brussels sprouts	Sown once, standing all summer; we sow Brussels sprouts two to three weeks before last spring frost in zone 4, transplant about four weeks later, harvest around five months later, after first fall frost and beyond

Broccoli and cauliflower are among the most satisfying to grow, an unforgettable culinary experience I hope you have or will savor homegrown. If nothing else, there is sheer beauty in observing how the heads form, monitoring their daily progress, marveling at how heat and rain expedite maturity. And then, finally, when the crop looks large enough, there's the pondering about when exactly is the right time to harvest.

Broccoli is a well-studied and well-developed group of cultivars with a wide breadth of offerings for succession planting. From super early maturing to heat tolerant to overwintering sprouting varieties, there's at least one season where a broccoli variety can slide into your garden, if not more. Cauliflower, in general, is less heat tolerant, and thus happier living in the margins where it's cooler. It offers crunchy curds to your plate late spring through early summer, and again in autumn, after first frosts have knocked down all the tender vegetables. Thanks to horticultural ingenuity, there are cauliflower varieties that are heat tolerant as well, though I've found this plant prefers the early and late season.

Spring, summer, and fall cabbages are a fixture in our garden. We grow mini cabbages for earliest harvests; red and savoy for summer eating; and all types for fall harvests and winter storage.

OPPOSITE
A trio of home-grown goodness in mid-June: Romanesco, Violetta, and Vitaverde.

Cabbages are my favorite form of garden art, especially red cabbages, with their contrasting venation, and the lily pad whimsy of their lower leaves. I have come to appreciate just how heat tolerant cabbages are, harvesting them in late May all the way through the sultry summer months, often continuously until sometime in November. (Like other brassicas, their flavor sweetens after a few good frosts.) In our zone 4 garden, that equates to seven months straight of fresh cabbages. That steady harvest is made possible by succession planting three to four times throughout the growing season; transplanting strong, four-week-old starts into the garden; and relying on plants with differentiated rates of maturation.

Midseason Successions: Turn Up the Heat

At the core of every home garden are many midseason succession vegetables. These foods don't take a full season to mature, yet they are not as fast as the quick successions I discussed previously. These foods are completely intolerant of the cool shoulder season, instead requiring warm soils and preferring long, hot days to truly thrive and produce abundantly—and in doing so, fill our plates, stock our freezers, and pack our jars longest of all. Their indeterminate productivity effortlessly inundates the garden, growing indefinitely until disease or

As the early cabbages and broccoli near maturity, and the radish successions have already been harvested, it's time to sow a planting of bush beans that will germinate and be ready to take over this bed in a few weeks' time.

frost catches up with them. Most of the vegetables in this section will begin producing food two to three months from sowing. Many can be direct sowed, and most provide extended harvest periods, producing again and again, unlike the shoulder season brassicas, most of which work hard to produce once.

If you had a successful spring garden, filled with your favorite things, consider these vegetables your second succession. It's a mental challenge to wait for ideal conditions to get this next phase of the garden established. I look to our soil to lead the way. Hot season crops won't thrive without proper soil temperatures.

About a month after our earliest radish, arugula, and bok choy harvests, we begin to direct seed our midseason successions for the hot summer. Our grandparents and other elders often share traditions of "plant the garden" holidays, like Mother's Day or Memorial Day, regardless of zone or seasonal fluctuations. These are one-day events that eventually feed us well later in the summer.

In reality, there is an expansive window of time to sow most of these seeds. Sow them either directly into your prepared garden, or sow indoors a few weeks ahead for transplanting, with the exception of the solanaceous crops (tomatoes, peppers, and eggplants). Most of these midseason vegetables are temporally flexible across the succession garden. As long as they have a good two to three months left in the season, they will charge ahead and work expeditiously at feeding you. If you miss the earliest time to sow, sow at the next best time. If it's something you love to eat, sow it more than once during the growing season, as space allows. This will keep plantings tidy, vigorous, and at the height of productivity.

Beans

It might just be childhood memories playing with my head, but a garden-fresh green bean on a hot summer day is sheer perfection. I love the familiar, scratchy pubescence that sticks to my tongue like Velcro®, and the ensuing vegetal aroma and sun-warmed sweetness that explodes in my mouth. It's a signal that the height of the growing season has arrived. As a bonus, it can happen with delightful haste with bush string beans, which produce in under two months when sowed at the right moment.

TYPES OF BEANS

Bean Type	Characteristics
Pole	Indeterminate growth from later-maturing beans that produce indefinitely, though more meager per harvest than bush beans; require support, so plan accordingly
Bush	Determinate growth, a new phenomenon in the long history of cultivated beans; produce for shorter duration (a few weeks) with larger yields per harvest

Bean Variety	Characteristics
Wax	Yellow type of snap or string beans
Filet	French-derived beans, round and delicate; harvested small
Romano	Flat-podded bean from Italy; variety of growth habits and colors, from green to purple to mottled
Soybeans (edamame)	Any garden with kids should include edamame; compact and highly productive; blanch and freeze for winter treats
Dry (shelling) beans	From garbanzo to pinto, bush to pole, dry beans make for plant-it-and-forget-it protein; left to dry, these beans generally take at least three months to fully mature; can be harvested and dried indoors if need be

The fastest way to take green bean harvests to new levels is to plant several different types of beans with different days to maturity. Beans evolved from a vining crop to exhibit a bushier habit as well, the latter of which produces in a shorter time frame. When you grow both bush and pole beans in different areas of your garden, you will yield a bumper crop from the bush beans while the pole beans trickle in. As bush beans start to fade, pole beans should arrive in full force, carrying your green bean fix for another month or more.

Before that inevitable fizzle, and as space allows, sow more bush beans as soon as the first succession starts to produce. Succession planting another round of beans keeps the harvest steady and strong, be it a handful of seeds in an opportune pocket of bare earth, or a few rows of seeds where perhaps your earlier greens just finished. The longer your growing season, the more latitude you have for succession planting beans to keep the harvests steady.

Pole beans are also great for succession planting. The only challenge is adequate vertical support for a second succession. In our garden, trellises are serious business, their location and plantings predetermined, and those plantings always end up being full season crops. Because I don't plant beans on my main season cucumber or butternut squash trellis, I am limited by my vertical space, and can only sow one succession of pole beans. I work around this constraint by interplanting a few varieties with varying days to maturity on the same trellis. That single planting is its own succession planting, naturally spreading harvests a little bit further, maximizing our chances of a steady stream of green beans from early July through September.

Bean varieties abound and provide an extended succession of nutrition. From left: Miles garbanzo beans; dry beans Tiger's Eye, Dapple Grey, and Speckled Cranberry; Toyha edamame; and bush beans Velour, Gold Rush, and Maxibel.

Sweet Corn

Summer isn't complete without fresh sweet corn, a popular midseason succession. It might surprise you to learn that corn seeds are surprisingly tolerant of cool soils, and can successfully germinate in cooler soils than most gardeners realize. Adventurous farmers with workable soil plant their corn seeds in late April here, a few weeks before our last frost. If you want the earliest sweet corn, you plant it among your earliest direct seeded crops.

Here in Minnesota, the fresh sweet corn season runs from around the Fourth of July to beyond Labor Day. Local farmers keep the supply fresh all season by succession planting it several times, and grow multiple varieties with varying days to maturity. In your home garden, a small planting block of corn sowed every few weeks, repeated two or three times as space allows, will keep a steady supply of super fresh corn coming for a month or more.

Cucumbers

One of the most prolific garden companions is a happy cucumber plant in late July and August. So happy, its favorite thing is to play hide and seek, and win day after day, resulting in enormous treasures when you finally get the angle just right and locate those rogue fruits. In addition to its productivity, this heat lover is a star in vertical gardens, its foliage thick and dark green, lush, and massive, expeditiously rising to the occasion, completely enveloping a garden trellis with adventitious shoots. Cucumber growth is most often short-vining, even when seed packages list it as non-vining bush type. Cucumbers originated in South Asia, and the greatest diversity of cucumbers can be found among Asian varieties.

Three short rows yield three family-of-four meals of roasted sweet corn in the summer, which is the just-right amount for a summer treat.

TYPES OF CUCUMBERS

Cucumber Type	Characteristics
Slicing	Most common homegrown type, noted for thicker skin and seeds; more bitter than seedless or English varieties; Marketmore76 and Poinset76 are both improved, open-pollinated varieties that produce well here
Pickling	Small, stout, highly productive, growing on shorter vines; stay crisp when pickled; smaller and mature more quickly; pick at any size to pickle or eat fresh; growing Sumter is our annual tradition
Seedless (burpless)	Parthenocarpic (self-fertile), so cross-pollination not required; result is thinner skin and seedless flesh, making these types sweeter than slicing cucumbers; examples include Persian, Seedless, and English types
Novelty (open-pollinated heirlooms)	Come in an array of sizes, shapes, and colors; examples include Boothby Blonde, cucamelon, Keera, Lemon, Suyo Long, and Armenian (actually a melon)

As with all vining crops, cucumber has the ability to send out suckers at every leaf node on the main stem. If left unchecked and unpruned (which I admit I did for the longest time), it can rapidly develop into an unruly mass, onerous to manage and unmotivating to harvest. To thwart this potential pitfall, I've developed a penchant for pruning all my vining crops. Vertically trellising nearly all of them makes for ergonomically friendly pruning conditions. I take special care to prune the cucumbers, but even things like butternut squash and cantaloupe get a good haircut a few times at the height of the growing season. It keeps them a little tidier and increases air circulation, and thus delays or eliminates the threat of diseases that capitalize on their dense foliage, such as powdery mildew—not to mention it's a tremendous help in eyeing ripe fruit.

Exploring new cucumbers each year has been a real joy. Our lineup from a few years ago included from the top: Lemon and Marketmore slicing; Sumter pickling and Persian; Armenian cucumber and cucamelons.

Planted two feet apart on each side of the trellis, I tamed these four lush cucumber plants by occasionally pruning their suckers, which thwarts disease by increasing air flow around the plants.

Summer Squash

To plant summer squash in your garden is to embrace summer's most authentic abundance. It begs you to find the joy and creativity in preparing the seemingly endless supply of this fast-growing vegetable. Be it a patty pan, crookneck, or traditional green zucchini, all are variants of the same species. As cabbages, tomatoes, and peppers come in all kinds of shapes and sizes, so too do squash.

Perhaps, like I sometimes do, you take a few days off harvesting, only to be met with a massive gift in the garden. Like with those sneaky cucumbers, your reward is an even greater amount of produce. I've quickly learned this unfortunate mishap can be avoided (mostly) by growing yellow or variegated varieties, which are easier to see. When in doubt, one summer squash plant will feed our family quite handily all summer long. Sometimes less is just right.

Not only are they massively productive, but the amount of space summer squash occupy at maturity is not to be underestimated. They quickly head for neighboring plants' plots if not given adequate room to roam. I tend to err on the "more is better" side with this vegetable, giving it at least a 4 ft. by 5 ft. area to

Highly productive and highly visible, I am loving the hybrid summer squash variety Goldy.

flourish, but I don't stop there. As with cucumbers, I prune my summer squash to keep them in check. It slows them down momentarily, and in that moment, we truly miss their presence in our primavera pastas and on our grilled veggie platters.

To my delight, summer squash has proven deer resistant, a requirement for food to thrive outside our deer fence, given the sizable herd that wanders our rural fringe neighborhood. Anecdotally, the pubescence of cucurbits is rather rough on the skin, as you probably know from reaching in for a squash or cucumber. Who would want to bite into that? It turns out deer don't find sandpaper-textured foliage all that palatable. This has been one of the easiest ways I've been able to add an extra succession of summer squash to our garden, by interplanting in a sunny spot, topped with compost, among our perennial flowers and other deer resistant vegetables like rhubarb, garlic, and onions.

Solanaceous Crops:
Tomatoes, Peppers, and Eggplants

The stars of most home gardens for a multitude of reasons (most notably flavor) are this group of prized fruits in the nightshade family we call vegetables. Tomatoes, peppers, and eggplants all produce plentifully midseason, thanks to some crafty forethought by either you or a local grower. These are all plants that do best when given a head start indoors, and are ready for life in the garden right around the time of your last spring frost.

Let's take the beloved homegrown tomato as an example. The fastest path for a tomato to go from seed to ripe fruit in my garden is three and a half months, and it is always a cherry tomato. It starts during my late winter indoor sowing, a distant daydream of sweltering July days with scattered thunderstorms, and ends with an early summer harvest, typically with an audible squeal. Tomatoes are not the fastest garden companions, but the advantage of this vining crop is that it produces for a few months straight or longer, if your season is mild and disease pressure is low.

Espalier Tomato Successions

Tomatoes are the succession garden we've been planting before we knew what that was. My family has always gravitated toward a diverse group of tomatoes that naturally ripen over a long period of time. I love how they trickle in weeks (and sometimes months) apart. We are partial to trellising our tomatoes like our fruit trees, in the espalier method (pictured). I train three main leaders from each plant, and spread them across the panels to provide air circulation and easy access to the ripe fruit. On a dry, sunny day each week, I take the time to prune suckers and tie down the leaders. With each plant 30 in. on center, it feels like plenty of room until about late July when the tomato bed is full of foliage and trusses of ripening tomatoes.

TYPES OF TOMATOES

Tomato Type	Characteristics
Indeterminate	Vining plant, growing indefinitely or until killed by frost or disease; ongoing flowering and fruiting once it begins; steady productivity, often over several months
Determinate	Finite growth period before fruiting; produces fruit more quickly and in larger quantities; once fruiting is done, plants will not produce anymore

Tomato Variety	Characteristics
Cherry	Small, round, sweet tomatoes, fast to mature; most often indeterminate
Grape	Shaped like a grape, hence its name; often smaller than cherry with thicker skin, thus more shelf stable; fast to mature; often indeterminate
Paste, plum, or sauce	Best for sauces or salsas; oblong in shape, meaty texture, usually dry with few seeds; midseason ripening, with both determinate and indeterminate varieties
Beefsteak or slicing	Large, juicy tomatoes, often growing in excess of one pound; latest to mature in most gardens, often indeterminate
Heirloom	Any type of tomato above can also be an heirloom, meaning an older, open-pollinated variety; examples include, Amish Paste, Black Cherry, Brandywine, Cherokee Purple, Yellow Pear

Tomatoes benefit from a few months of initial growth indoors to establish and prepare for setting fruit. Don't underestimate the energy required for tending them inside before transplanting. We have started them as early as February 1 and as late as April 1, and while the February plants require a lot of potting up—eventually needing 1-gal. pots and clinging to our sunniest windowsill because they've outgrown our plant stands—the extraordinary effort always pays off in very early cherry tomatoes. We reliably harvest the first of these in late spring, a week or more before summer solstice, a remarkable feat that thrills and fuels our passion for season-extending.

Started on the same February day, by April you can see how much faster tomatoes establish than peppers.

For peppers and eggplants, the wait is usually a good bit longer, but so very worth it. These plants are even slower to establish from seed, almost painfully so, especially for new gardeners. Plants started on the same day quickly look like two separate successions, the tomatoes' more vigorous habit swiftly towering over the slow but steady establishment of the peppers and eggplants. It can be a

Tomatillos and Ground Cherries

Another subset of solanaceous plants that produce in this middle succession at the height of summer is tomatillos and ground cherries. Both plants set husked fruits and have potato-like foliage. Tomatillos are most commonly used for roasted sauces and salsas, while ground cherries are a novel garden fruit, sweet and distinct. Like the rest of this plant family, these also hail from the Americas—some are even native to North America, and for some reason it intrigues and delights me to grow an annual food that's evolved over millennia on the very continent we make our home. Like many tomatoes, they are prolific producers, and readily self-seed.

Tomatillos and ground cherries are the only solanaceous food I would ever direct sow. But actually, it's never me who sows them. Flaunting their unfettered will to thrive, these plants never fail to sprout in far flung locations around our property, a nod to the bird or rodent who enjoyed them as much as we did the previous year. While I don't rely on found volunteer tomatoes or peppers to become my primary crop, I would for these plants, for they tend to be leggy and floppy as indoor starts, even in ideal growing conditions. Meanwhile, the self-seeded volunteers are sturdy. While they germinate a little behind the indoor sown plants, by the end of summer, there's little difference between the two.

mental challenge to trust the process, but eventually they will take off.

Similar to cherry tomatoes, hot peppers are the earliest to produce, a boon to short-season growers and pepper lovers alike. Ripe bell peppers don't usually arrive until a month or more after our earliest small pepper varieties, as with beefsteak and cherry tomatoes. Because the season for peppers is much shorter than it is for tomatoes, they really are a delicacy of the late summer garden, and planting a wide variety is a great way to add successions within successions to your food garden.

While these all take a good four months to produce from seed, I consider them midseason succession crops because of their frost intolerance and dislike of the cool soils of the shoulder seasons. Timing their establishment in your garden so they're strong and healthy as soon as environmental conditions are favorable—which is usually one to two weeks after last frost—will result in the longest season possible. And for those who endure long winters, ripening tomatoes and fresh peppers are among the most anticipated homegrown foods to mature.

A Finicky Midseason Succession: The Elusive Garden Pea

In addition to so many heat-loving crops in the midseason succession, other, relatively early season crops like peas also require a full two to three months to mature, including snap, snow, and shelling peas. Timing peas can be downright tricky in many climates, filled with uncertain success at the mercy of early season weather. A quick warmup in spring (often the case in many northern climates) means less than ideal growing conditions for these true cool-season lovers as they head into flowering and pod production. Sowing and establishing them during the heat of summer for a fall crop has repeatedly let us down, neither yielding well nor holding up to fall frosts quite like their spring siblings do.

To get ahead of our notoriously finicky spring weather, we plant peas in the ground as early as possible so they can settle in and thrive by June, usually producing two to three weeks from mid-June through early July. Season extending methods combined with pre-soaking the seeds result in pea shoots popping out of the chilly earth before the middle of April. Some years are better than others, both in productivity and timing.

Extra early sowing of spring peas guarantees a healthy harvest by the end of June, and goes on to provide a welcome gap in our summer garden for a succession of carrots, beets, or fall cabbages.

A key lesson I have learned is that peas enjoy being crowded. So, go on, plant densely, and don't thin them out. Peas like to grow close together, and anecdotally, are more productive the more densely they're planted. It's a practice in flexibility for me because I err on the side of more space being better, except in this case.

Late Season Successions

Some of the sturdiest, most shelf-stable vegetables are downright space hogs, taking a full growing season to mature. Yet, they are worth the space, because they patiently wait for you, don't require processing to be stored, and many are disease-resistant (and thus reliable) crops for home gardeners. They will last for months in proper storage, providing self-sufficiency as well as deep homegrown pride that your potatoes, onions, popcorn, and squash will get you through the holidays and beyond.

Now most of these foods take a full season to produce. And by that, I mean at least four months from seeding to maturity. Other vegetables covered in this category are most often grown as late season successions because they grow best in the cooler, softer days of the fall garden. But don't be fooled by their love of cooler weather. In order to succeed in fall, many of these plants must be started during the heatwaves of summer—or sooner.

Popcorn and dent corn, pictured here, nourish us during the slower and colder months, while sweet corn is relished in the heat of the moment.

Flint, Dent, and Popping Corn

Sweet corn aside, other varieties generally require significantly more days to mature. This is because unlike sweet corn, which is harvested in the early stage of maturity, popcorn, flint, and dent corn are all harvested in later stages. We use their kernels dry, so they must remain on the plant for more of the growing season to fully mature. While sweet corn grows to about 5 ft., other varieties tend to grow 7 ft. to 11 ft. This makes them a great living screen, and structure for climbing crops like beans.

Growing pantry staples like popping corn and dent corn for cornmeal is a luxury afforded by our rambling growing space. These varieties anchor our garden quickly and sustain themselves for the duration of our frost-free growing season. I have come to adore their presence in our garden, their stature a statement that always draws me in. From variegated leaves, to gem-like ornamental kernels, to petite plants with petite popping corn cobs, a small bed of these pantry pleasers will be sure to delight you in the winter months. The corn garden sustains in more than one way.

TYPES OF DRY CORN

Corn	Characteristics
Popcorn	Dry corn popped for eating; one of the oldest types in cultivation
Flint	Hard outer layer similar to popcorn, this type of corn may be popped, but pops inconsistently; often grown as ornamental corn; takes over four months to mature; allow to mature on stalk as long as possible when weather is favorably dry
Dent (also called field corn)	Named for dent kernels that develop as they dry; most commonly grown for cornmeal, tortillas, taco shells, ethanol, and livestock feed

Potatoes

I consider potatoes a full-season crop, and a late succession one at that, in part because of how and why we grow potatoes. Potatoes are one of the cornerstones of our root cellar. We grow this vegetable to sustain us through winter. We grow just one succession annually, though it can certainly be succession planted, set out across a few weeks in early spring and summer, especially if the varieties you choose are for fresh eating. Growing a few different varieties that mature at different times will naturally yield a steady stream, plus some to store for the colder season.

Because it is a staple of our winter diets, we plant potatoes around the traditional "plant the garden" holiday, typically Mother's Day here. We generally leave them in the ground throughout the growing season, letting the plants fully die back, waiting to harvest the majority of this beloved solanaceous tuber until mid- to late September. This strategy is in large part because space constraints are low, and we use our garden as a root cellar of sorts. It helps that the food is quite content to wait to be unearthed, remaining in peak condition below ground until we are ready to harvest.

Like with peppers and tomatoes, potatoes offer a wide range of maturation times. Potato lovers can strategically plant an early variety for fresh eating, and a main season or late season variety for storage. This is especially useful if you're growing in a small urban garden where space is at a premium. Potatoes also grow very happily in containers or grow bags, creative ways to add grow space to your home garden.

AmaRosa harvest in late September is a joy to unearth—1 lb. of seed potatoes became a 55 lb. harvest. A meditative and messy process, but one that feeds us for months.

TYPES OF POTATOES

Potato Type	Days to Maturity	Common Varieties
Early season	55–75 days	Caribé, Norland, Yukon Gold
Mid– to main season	80–100 days	AmaRosa, Charlotte, Kennebec, Magic Molly, Red Pontiac, Rose Finn
Late season	90–110 plus days	Butte, Carola, German Butterball, Russet, Russian Banana

Winter Squash

It is a gift of nature that foods vary so much in days to maturity. This is one of the many keys to succession gardening, dialing in these nuances and planting a diverse lineup so that your food doesn't arrive all in a few weeks' time, but rather succeeds in waves; multiple, delicious waves that sustain you for as many months as possible. Winter squash is a favorite plant of mine in this regard. Plant it once, in June, give it a structure to maximize garden space, and let it be. Winter squash produces bountifully, is more resistant to the pernicious squash vine borer, and one of the easiest vegetables to store, naturally arriving with shelf-stable packaging that requires little fuss. Winter squash stores well for several months in a dark closet.

I consider winter squash late season successions because they take a full season of growth to mature, about four months from seeding. As with potatoes, we generally don't harvest them until very late in the season—but before hard frost, which dramatically deteriorates storage quality. Thus, the space we allot to winter squash is a single succession per year. For me, that means I don't grow anything in its spot before this food goes in, and I also don't grow anything there after it's harvested.

Alliums: Onions, Shallots, Leeks, and Garlic

Late succession foods and culinary staples, onions, shallots, leeks, and garlic take several months to reach maturity. Despite their harvest in late summer, when many mid-succession crops are also in season, these vegetables require a full six months or more to reach maturity. Onions , shallots, and leeks are seeded indoors in mid-February through mid-March, tended and trimmed in their crammed 4-in. pots, and transplanted out to the garden around May 1. Onions and shallots come out of the garden about three months later, a full six months from seeding.

Two Waltham butternut vines cling to the support of cattle panels, while the squash themselves cling without support until harvest time in early autumn.

From seed is the only way I've ever grown onions. They are economical and don't mind being crammed together temporarily (saving precious grow light real estate) until they are hardened off and ready to transplant a week or so before our last spring frost.

Garlic harvest in late July creates a gap for a fall garden succession to quickly fill in the space.

Leeks, however, stay in our garden as long as possible, blissfully neglected and patiently waiting for a late fall lifting from garden to root cellar.

Garlic is a sturdy and fascinating late succession food. Garlic cloves get planted in late October, miraculously lying dormant after setting their initial roots, alongside other hardy perennial foods and plants, emerging as a delightful surprise early the next spring. Hardneck garlic reliably weathers one or more spring snowstorms and goes on to produce a bonus harvest—a flower head known as the garlic scape—in late June. About a month later, the plants themselves are ready for harvest.

Garlic and onions are integral crops for a succession garden. They mature just in time (late July to early August) to set out strong fall starts immediately after their harvest. Space that opens up here six to eight weeks before the first fall frost begs to be filled with our favorite fall foods. That way, we know we've maximized our season, while also having reaped some shelf-stable food for the fall and winter months too.

Brussels Sprouts

Brussels sprouts win the crown as the only brassica that takes a full season to mature—and I mean all season long. There are slow growing cabbages too, but Brussels sprouts occupy space in our garden from late May until late November. Yes, that is a remarkable half-year, and longer than our frost-free growing season. Their presence extends well beyond our first fall frost, and out past several hard frosts. They are a uniquely hardy brassica dating back hundreds of years, another modified leafy portion of the same species of *B. oleracea*, which also produces broccoli, cabbage, cauliflower, kale, collards, and kohlrabi.

Late season Brussels sprouts mingle with quick and midseason succession vegetables, providing structure throughout the growing season.

My fascination with this crop is its time to maturity relative to the diminutive stature of its produce. After all those months of photosynthesis and anticipation, it yields several dozen mini, sweet cabbages at every leaf node of the stem. It's ironic that more pounds per square foot are produced faster from a cabbage. However, Brussels sprouts provide diversity; their slow maturation is a boon to a succession garden because it makes them a reliable late season food that you can truly set and forget, aside from checking for pesky imported cabbageworms and cabbage loopers. Otherwise, you can simply relax, marvel, and wait—not to mention the fact that their extreme cold tolerance means they can remain in the garden months after the first autumn frost.

I find ease and comfort devoting a significant portion of the summer garden to late succession foods—foods that are familiar and sought after in our kitchen adventures, reliably productive and shelf stable, and that will wait for me rather than demand my immediate attention. While the mid-succession heat lovers fill my calendar with processing tasks, late season successions just plod along, seemingly unhurried, while the garden all around them erupts in a sense of immediacy. Some days, I wish I could channel the zen-like growth of my late succession crops amid my cucumber hunting, tomato canning, and Japanese beetle flicking.

Generalists: Anytime Mid-Succession Vegetables

There's a useful group of vegetables that don't mind cold or heat. These can be sown under a row cover in late winter, or in the middle of a summer heat wave. They sail through the seasons like the sturdy, dependable garden companions they are. We lean into these vegetables often to ensure our garden is in peak diversity and production throughout the growing season. They mature in the sweet spot, two to three months after planting, and also weather seasonal fluctuations as if they were made for every single one of them. These vegetables were made for the succession garden.

Beets

Beets are my kind of generalist succession. In our short but sweet growing season, I can sow or transplant beets practically any time and they produce for me. Lean into beets for that added diversity and dependability. Yes, you must embrace the earthy tones of a root you (like me) may have spent your childhood avoiding. Yet like so many other vegetables, the entire plant is edible. The roots store well in the fridge for several weeks in summer, and for months tucked into our root cellar. Likewise, roasted beets add complex flavors to a green salad in summer, and accompany roasted squash and turnips perfectly in fall dishes. And don't forget to enjoy the greens too.

It might surprise you to learn that beets grow well from transplants. What shouldn't surprise you is that this is a bit of a controversial approach, being that beets are a root vegetable—which generally speaking, are always direct sowed into the garden so as to not disturb their precious root systems. I stumbled upon this transplanting concept by accident after misremembering a conversation with an organic farmer.

Turns out our farmer friend doesn't indoor sow their beets, but now that is our sole method of growing them. Like with most other crops, germination is faster and more consistent in a controlled environment. And because beets are an aggregate seed, what we consider a beet seed is actually a cluster of seeds that germinates multiple plants, I am able to easily tease apart their clustered seedlings in our soil blocks and space my beet planting in perfect rows when I transplant. They also mature more quickly when sown indoors and transplanted because of the ideal conditions they enjoyed the first month of their lives.

I sow my beets in 2-in. soil blocks and thin them out carefully when transplanting around five weeks from the date sowed.

Carrots

If there is one vegetable I wish everyone would grow for themselves at least once, it's the garden carrot. From the sheer delight and awe in harvesting them to their fresh crunch and sweetness, which store-bought carrots lack, a home-grown carrot is a transformative culinary experience. There are the carrot years before homegrown, and then there's life after eating a homegrown carrot.

I place carrots right up there with tomatoes as non-negotiable annual garden vegetables. The flavor is worth what sometimes seems like the herculean effort to successfully germinate the seeds, and the subsequent months of tending and waiting for those tap roots to swell. If you're lucky enough to live in a cooler climate, early fall frost signals this biennial crop to accumulate sugars in its roots, storing up energy for next year's seed production. It also means sweeter carrots for all to enjoy. We prefer to wait to harvest the bulk of our storage carrots until after a few hard frosts.

Carrots are a midseason succession that can be added any time of the growing season. Just remember to add them at least two to three months before your first fall frost, so you will be rewarded with garden fresh candy straight from the soil. We enjoy sowing carrots all season long, starting early and maintaining a sowing mindset. As we harvest our spring garden and create gaps, we mostly sow with carrots for fall. It's our favorite summer succession to add.

We exclusively sow our carrots under a damp single layer of burlap. The fabric holds in added moisture, keeping moisture levels even at the soil, and hastening germination. Once the carrots germinate, it's easy to lift the single layer off the plants without disturbing their root systems.

A medley of carrot types makes for a diverse harvest. From left to right: Atlas Parisian carrots, Danvers126, Yaya, and Purple Haze.

Regardless of soil type, there is a carrot for every garden. Clay soils or container gardeners might lean more toward the Parisian type, a round and sturdy carrot that bulbs even in the most challenging soils. Cold season gardeners might prefer a Nantes or Chantenay type, prized for storage quality, and becoming sweeter after those first few frosts. And if you're after the rainbow, there's a packet of carrot seeds just for that too.

VARIETIES OF CARROTS

Variety	Use	Characteristics
Chantenay or Oxheart	Heavy clay soils	Shorter, more stalky roots; varieties include Hercules, Red Core, and Short Stuff
Nantes	Fresh eating	Smaller, cylindrical carrot with blunt tip; Yaya is our favorite Nantes variety; also grown for storage
Imperator	Storage	Well known grocery store carrot, long and skinny; cross between Nantes and Chantenay; requires loose soil to perform well; varieties include Sugarsnax
Danvers	Fresh; short-term storage	Broad shoulders with strong taper; developed in Massachusetts; Danvers and Danvers126 are varieties

Back to Brassicas

I would be remiss if I didn't round back to this robust group, acknowledging their rightful place as a garden generalist, particularly in our northern garden. Kale and broccoli are probably the most notable and sturdy of them all, though cabbage also fits the bill in our home garden, producing in about three months' time. Our garden is never without a planting (or five) of brassicas, for their maturity rates vary so much that they provide a beautiful succession of nutrition throughout the growing season and beyond.

Basil is an easy beginner gardener herb to grow from seed. Left to right are some of my favorite varieties: Genovese, Tulsi, Opal, and Thai.

Herbs

Herbs are a diverse group of plants I also consider generalists, and gateway garden plants for all the best reasons. They grow well across a wide range of the growing season. They are quick to mature, providing near-instant gratification. And you can sow multiple successions every few weeks during the growing season for a steady stream of flavor boosters. Most annual herbs are mid-succession crops, maturing around two months from sowing.

Herbs take any homecooked meal to the next level. They grow well just about anywhere: in pots, windowsills, balconies, or in the ground. Give them sunlight, nutritious soil, and not too much water—herbs are generally more drought tolerant than vegetables—and you'll be infusing your meals with a burst of flavor in no time.

Best Methods for Starting Herbs

Some herbs sprout readily from seed, while others grow best from cuttings. Most herbs that grow readily from seed are fast-growing annuals:

- **Basil**
- **Chives (perennial)**
- **Cilantro**
- **Dill**
- **Fennel**
- **Parsley**
- **Thyme**

Most that grow best from cuttings are shrubby, and largely perennial:

- **Lavender**
- **Lemon verbena**
- **Mint**
- **Oregano**
- **Rosemary**
- **Sage**

The beauty of understanding and mastering vegetables for succession is that you can then personalize your garden across the four seasons. Fine-tuning your plan, selecting the best seeds or starts, mastering the timing, and playing with your space, you make the most of each and every growing day of the year. Creating layers of vegetable succession in your garden is an act of constant renewal, of actively being in relationship with your land, of sowing hope for a fruitful and delicious tomorrow. These layers are so much more than just food on the table. They reduce your carbon footprint; provide access to fresh food; and encourage you to connect with nature and its intrinsically healing modalities on a daily basis. It is a conscious lifestyle that rewards those who stay present and active in their garden planning.

The beauty of a succession garden in our climate is the raucous display of diversity come August.

A monarch magnet, meadow blazing star resides in both our food and auxiliary gardens, so we can commune with these butterflies as often as possible in summer.

FLOWER POWER

Flowers are integral on the landscape of the most productive food gardens. The wise gardener sprinkles, showers, and supports their food garden with flowers around, among, and adjacent to their most desirable edible crops. Planning for a succession of nectar and pollen across your entire growing season ensures robust habitat for the essential workers who pollinate as much as one third of the food we eat.

From lawn to vegetable garden to formal garden, there are truly endless permutations of opportunities to personalize your garden as a welcoming habitat for insects. In doing so, you transform these areas into pollinator havens. And let's just be clear: gardens succeed because of insects. Their ecological significance has taught me to invite them in, to stop and marvel, and to work to protect their habitat on our landscape. We need each other to succeed.

I am a bit of a control freak, and my carefully curated food garden is no exception. But gardening for pollinators is an exercise in anything but wielding control over nature. It's quite the opposite, and welcomes what might look like chaos in your neighborhood—and if your neighborhood is anything like ours, it is carpeted with acres upon acres of sterile monocultures in the form of turfgrass, which you've been conditioned to love through generations of advertising and marketing.

I have learned over the years through a slow, steady accumulation of observations that a wild and unkempt area of my former lawn is the healthiest, most diverse patch of land on our landscape. And its diversity has tremendously positive downstream effects on my food garden, providing habitat and forage for both predatory insects and native pollinators, the very guests whose arrival coincide with important spring flushes, like our fruit orchard blossoms. Providing habitat for these insects ensures we and they are each provided with the food we both need and love.

A simple way to provide habitat for native insects is to leave an area of your property unmown. An undisturbed area is a signal of safe haven for ground-nesting bees, which make up a remarkably large percentage of native bees. Ground-nesting bees require this diminishing and severely fragmented habitat to successfully reproduce and pollinate our prized fruit. And if you aren't going to mow an area, well, that's simply a perennial flower garden begging to be planted.

Plants are rich and alive with memories, their presence holding space for people and places that may have passed. Such is this remnant prairie, a permanent nod to the former homes we've inhabited, now commingling here in spirit.

That's just what my family has done in every home we've owned, taking out turf and adding native perennials, primarily because we reside on the edge of a historic tallgrass prairie ecosystem. For a few short and storied years, we were fortunate enough to own land with about 3 acres of planted prairie. It was my first foray into living alongside prairies on a larger scale, across the seasons, and it indelibly shaped my relationship with the land in the Midwest. Not only has growing food become a vocation, but restoring portions of land to native plant populations has also become a key component of our landscape plans.

The Garden Buffer:
A Case for Auxiliary Flower Gardens

Ringing our food garden is a massive buffet of nectar and habitat. About three times the area of the actual vegetable garden, our auxiliary gardens bloom across the seasons, and the majority have been planted as native prairies. This is as much for our personal enjoyment as it is for the benefit of our ecosystem.

Inherent in this process of working with nature is that we all derive joy from working toward the same goal. For humans, it's the sought-after color and beauty that a succession of flowers provides all season long. For insects, those same flowers are their lifeblood: essential sources of nutrition, beacons of seasonal forage that help them take notice of other such blossoms, like tomatoes and squash. When we witnessed how pioneer prairie flowers like black-eyed Susan and early figwort attracted and delighted the bumblebees, it was clear our landscape was on the right track.

The auxiliary garden is one of the simplest and most important facets of our personal gardening success. This naturalized garden emulates what we see nature do everywhere. Nature is the interplanting, succession-planting expert, mixing together plants who mature across the seasons in one small area, so that there is always something for wildlife and pollinators. And maintaining a piece of land with a similar ebb and flow is a powerfully instructive model for how I approach succession planting.

Monarch butterflies are another reason I love our auxiliary gardens, as they provide critical habitat and forage to these threatened long-distance travelers.

Just as we would like to eat for as long as possible from our gardens, so too would the insects, and planting various flowers that bloom from early May all the way through October provides native pollinators an ideal habitat. In the case of native bees—the bees whose growing numbers on our property I'm most excited to observe—early, mid-, and late season flowers provide critical habitat. Native bees' life cycles span the growing season, and extending the season of feasting ensures that all types of these solitary ecological warriors can thrive. While some specialized pollinators like mason bees emerge in late April, others don't hit their stride until July or August, including certain cuckoo, leafcutter, and sweat bees. Many of the early season species are critical for successful pollination of fruit trees and berries. The success and wellbeing of these species brings literal sweetness to our lives.

More than enough pollen and nectar from our auxiliary gardens means that not only can populations of introduced honeybees thrive, but so can native bee populations. Honeybees are known to outcompete native bees, because they fly longer distances and their pollen sacks act like tightly-sealed baggies. Incredibly, mason bees only fly 300 ft. from where they create their nests. The vast majority of pollen collected by honeybees is tidily carried back to their hives, rather than sloppily dropped here and there, as native bees so generously do. I appreciate seeing all types of bees here. The diversity of bees and pollinators signals a diverse and plentiful feast of nectar and pollen across the seasons.

Bumblebees

If I had to pick a single species of native bee that most appreciates the targeted efforts of gardeners, it would be the bumblebee. A social bee, like the honeybee, this insect is a major garden friend. Pollinating most fruit trees, edible shrubs, tomatoes, peppers, and eggplants, the bumblebee's specialized dedication to a single flower type (called floral constancy) means it frequents one kind of flower at a time, and thus is a supremely effective pollinator. While this hyperfocus may lead to cross-pollination, many plants wouldn't produce well without its focused efforts. The bumblebee's larger stature means easy access to more nectar than smaller bees, because it is able to pry open bean, pea, and snapdragon flowers. Add in its *sonication*, or buzz pollination, and long tongue, and this generalist bumps elbows with the efficient honeybee. It's a superhero of the food garden. While bumblebees' penchant for cross-pollination might not be ideal for your tomato seed–saving plans, your fruit orchard wouldn't be nearly as successful in their absence.

Anise hyssop is an early flowering native perennial and summer magnet for pollinators, especially bumblebees.

Again, adding as many different types of flowers to your garden in as many different seasons as possible is the key to unwavering pollination rates and success. Species native to your area have evolved to keep their flowers blooming and fruit swelling, and you can further support pollination by adding a few of your native plant species. This holds true for both shady and sunny gardens. Here are some stellar native perennials for native bees to add to your landscape.

NATIVE PERENNIALS ACROSS THE SEASON

Perennial	Bloom Period	Native Range	Light Requirements	USDA Hardiness Zones
Golden Alexanders (*Zizia* spp.)	May–June	Varies by species	Full sun, partial sun	3–8
Beardtongue (*Penstemon* spp.)	June–July	Varies by species	Full sun, partial sun	3–8
Milkweed (*Asclepias* spp.)	July–August	Varies by species	Full sun, partial sun	3–10
Common Yarrow (*Achillea millefolium*)	June–August	Throughout North America	Full sun, partial sun	3–9
Wild Bergamot (*Monarda fistulosa*)	Late June–September	Throughout North America except Oregon, Nevada, California	Full sun, partial sun	3–9
Mountain Mint (*Pycnanthemum virginianum*)	June–September	Eastern United States through Great Plains	Full sun, partial sun	3–7
Blue vervain (*Verbena hastata*)	July–September	Throughout North America	Full sun, partial sun	3–8
Sweet Joe-Pye weed (*Eutrochium purpureum*)	July–September	Eastern United States through Great Plains	Partial to full shade	3–8
Sneezeweed (*Helenium autumnale*)	August–October	Throughout North America	Full to partial sun	3–9
Goldenrod (*Solidago* spp.)	July–October	Varies by species	Full sun, partial sun	3–8
Aster (*Symphyotrichum* spp.)	August–October	Varies by species	Full sun, partial sun	3–9

Every year we observe new and fascinating insect species in various stages of development in our garden. One spring, viceroy caterpillars, with their intense biomimicry, repeatedly found their way to our Haralson apple tree. These larvae generally feed on willow and related tree species as their host plants, so our curiosity was piqued when they took up residence on the apples. Were they coming to pupate? The classroom aspect of keeping wild patches of land—places where nature can thrive—is a perennial invitation to be present and learn, and to be humbled and inspired by the wonders of the highly orchestrated natural world. I lean into the knowing of the creatures there, and always feel a surge of joy at their presence in our garden.

Sometimes all that is asked of us is to listen to what is right before our eyes. When we first found that larva, we removed it, concerned about potential foliar damage to our apple tree. We placed it somewhere safe so I could keep an eye on it. Next, I photographed and worked to identify it. We wanted to understand what species it was before making any decisions. Once we identified it as a viceroy, we placed it on our weeping willow tree, its larval host plant, and several weeks later observed a few viceroy butterflies mingling about the auxiliary pollinator gardens.

Viceroy caterpillars feed on willow species, so it was a mystery to us why this one kept finding its way to our Haralson apple tree.

Living Lawns

Most homeowners aren't giving up their mowed lawns. They're a core cultural value of what it means to own property and tend a home. But what about the possibility of infusing your lawn with flowering plants such as nitrogen-fixing clover, and transforming an ecological dead zone into a living habitat? The most asked-about design element of our food garden is our living lawns, and I see the appeal.

As an expansive green lawn appeases our eyes, so too does a tidy food garden, complete with raised beds and some type of mulch. We value a clean aesthetic and design in home gardens, probably because it seems easy to maintain. However, on the scale at which we're currently gardening, I felt that maintaining so much open space would demand unnecessary weeding hours, maintenance, and resources.

So, we thought to ourselves, why remove existing vegetation only to create a dead zone with mulch or pea gravel? What if our paths could remain living, provide added color and visual interest, and more importantly, pollinator habitat on the landscape? Granted, our climate is prone to increasingly frequent

A living lawn is so simple and effective—and for us, a lush green carpet all summer long, more than compensating for all those months of muted winter hues.

flash floods in summer, and thus naturally well-irrigated, so maintaining lush, green paths throughout the growing season is about as easy as remembering to breathe. I would be remiss if I didn't acknowledge that our climate works to our benefit for this approach.

Like a beautiful specimen tree, our gardening style has taken decades to fully mature. Informed and influenced by our local climate and available resources, we've slowly amassed knowledge and shifted our ideals and preferences. In planning our current garden, we incorporated living paths in the design for the first time, and I love that our garden paths have been transformed from an insect food desert into a robust cornucopia.

The paths mean there is maintenance involved in keeping our in-ground beds edged, but their seasonal tending is rewarded with time spent reveling in the beauty of the garden at ground level. Distracted but inspired by frequent fly-bys and buzzing, it's an opportunity to slow down and just observe the

Mowed paths inside our food garden provide yet another layer of nectar succession, bringing joy and nutrition to honeybees and native bees alike.

immeasurable beauty of nature. If we weeded the area regularly because it was mulched or graveled, it wouldn't be nearly as inviting to kneel on as the soft, comfortable bed of clover. Above all, we simply love how beautiful living paths tie the garden together.

Dutch white clover, overseeded throughout our entire 2¾-acre property, brings flowers to our lawn, and overseeding with clover has become a hallmark of our landscape design. Long before bee lawns were a thing, nitrogen-fixing clover was ubiquitous in lawn seed mixes, a throwback to pre-World War II herbicide- and chemical-free lawns. It is soft, stays green, flowers multiple times throughout the season, and is a playground for four-leaf clover hunts.

Living lawns add yet another succession of nectar for those crucial pollinators as they flower several times across the growing season. It is space otherwise underutilized by insects, flyover land for hardworking bees in search of food to nourish their broods. Converting some or all of your traditional lawn or garden paths into a flowering lawn makes a beacon of hope between otherwise fragmented habitats. Creating a corridor of support and refuge for these critical ecological warriors couldn't be easier.

The Case for Natives

While not all of the flowers in our garden are native, or even "straight native" species that aren't human-bred cultivars, there is no better time than right now to choose native species over cultivars for gardens. These plants serve your native insect population with the highest quality nutrition they've known and loved since long before European settlement.

My heart has been leading me toward this philosophy for many years. I so appreciate the floppy, wild, and knowing way of a native planted prairie. I know that when I simply add native seeds or plants—those evolved to be most compatible with local native insects—I am providing the highest quality habitat possible. Because my fruit orchard hinges on the native bee population, gardening to support that population is paramount. We select nearly all straight native seed or plants when adding native perennials to our property, and largely avoid cultivars.

I consider cultivars "native-light." Given the choice between a lawn and a planting of cultivars, the latter is clearly superior—there's a beneficial gradient

between native plants and their cultivated relatives. Cultivars were native flowering plants many plant generations ago, and have since been bred for specific traits. Cultivars are often bred for more elaborate flowering parts under the careful direction of humans, and as a result, their interspecies relationships are no longer prioritized, and often rendered less useful.

Bred for characteristics such as color variety and longevity, which are pleasing to consumers, the resultant flowering parts are often difficult for pollinators to access. As such, their ecological usefulness (relative to straight native species) continues to be studied. And, confounding matters, their pollen is potentially a threat to native populations if (or more aptly when) it returns to wild populations by wind or insect, cross-pollinating with native species.

Cultivars of coneflower and black-eyed Susan are two of the most common found in Midwestern gardens. Both derived from very successful native plants, these summer dazzlers are a favorite of native insects. Coneflower seed heads provide forage for goldfinches in fall and winter, their purpose reaching well beyond simply feeding insects and our own visual delight. If you're planting straight native species, you guarantee the nutritional value wildlife rely on. As desirable as cultivars with sturdy stems or more elaborate blooms may be, when adding perennials, you should sincerely consider planting at least a portion of your landscape in straight native species instead.

Pale purple coneflower is a delicate native echinacea of the tallgrass prairie whose fleeting inflorescence never goes unnoticed by early summer bees like this green sweat bee.

Distinguishing Between Wildflowers and Native Flowers

It can be overwhelming to identify which seed mixes contain straight native plants and which ones don't. Even knowing the difference, I find this task confusing—and more than a little misleading. I have learned to proceed with caution if I see a packet of seeds labeled "wildflower mix." The term wildflower is widely used to market self-seeding flowers, both annuals and perennials, regardless of their place of origin.

Wildflower mixes usually blend native and non-native flower species unless otherwise specified. They may include old-time favorites such as oxeye daisy, which is considered invasive in most states because it displaces native plant populations. The annual flowers in these seed mixes readily self-seed, which delights consumers early on. This early reward is designed to help eager gardeners muster the patience to wait until the full spectrum of wildflowers in the mix bloom, because native wildflowers take anywhere from one to ten years to flower after seeding. Many take two to four years, but the really spectacular species will keep you on the edge of your seat for closer to a decade. Can you imagine the inquires seed companies would field from impatient customers as to why their wildflower mix isn't flowering if they had to wait more than a year for blooms?

If you're going for an ecologically potent flower garden for your native pollinators, be an informed consumer. Do your due diligence before purchasing seed or plants. I've learned over the years to thoroughly research plant selections ahead of shopping so I can successfully source what I need. At first, when gardening for bees and pollinators, I admit I haphazardly added any flower, I was just so excited for the variety of color and texture. Now that I focus on native plants, I almost exclusively seed shop for my perennial flower gardens through native seed catalogs.

I often do this year-round, at my leisure, so I largely avoid the pitfalls of impulse purchases—though even with this level of intent, an extra seed packet or two inevitably ends up in my cart. Native seed companies usually sell plant starts as plugs too, so you can invest more and start with plants that will flower

When the prairie to starts to flower, the insects arrive. Soldier beetle larvae prey on grubs, and I'm hoping they have an acquired taste for Japanese beetle larvae.

sooner, although seed remains the best economic value. The better informed you are, the easier and more confident your decisions will be in selecting the best possible plants for your landscape.

Be sure to do your research on what plants are native to your area, or close to where you live. These plants will find suitable habitat in your home garden. Native plants grow well in a range of conditions, so consider your soil, water, and sun as you dream up these gardens. Use native seed company websites, USDA plant databases, and the websites of well-respected arboreta to research new species. With climate change, plants are likely to migrate north, so consider the long view as you plan out your perennial auxiliary garden and set it up to be as resilient and successful as possible in the long run. Even the addition of a small number of key native species will have a tremendously positive impact on your garden and your local insect population. If you're converting an area of turf to flowerbed, so much the better. When planning for a pollinator garden, keep in mind that it will hopefully become home to ground-nesting insects, which seek out bare patches of earth to lay their eggs in broods. Because many native bees require bare ground to successfully nest, be sure to keep at least some of the surrounding soil exposed (not mulched with wood chips), or select compost as your mulch.

A view in the third year after seeding and the first year not mowing our native planted prairie. We enjoyed dozens of flowering perennials—and the insects abounded with all they offered.

Native perennial seed is the most economical way to add large perennial gardens to your landscape. Most of these plants' seed require stratification to break dormancy. Their internal clock prevents them from germinating until conditions are appropriate—it protects the seed in the fall from germinating before winter, when their young root systems would be damaged by the plummeting mercury. It definitely takes more time to establish a perennial garden from seed, but the journey is rewarding.

I remember vividly watching as pioneer flowering species took root and charged forward, surprising us with flowers in the first year of growth. Planted in a sprawling, permanent space on our property, we hope these and other long-lived species will be with us for decades, just like our trees. As such, we have given them a dedicated, sunny home where I won't be digging them up. I plan to let them root deeply, and enjoy their succession of nectar and pollen through many growing seasons. Each year is met with opportunities to learn how to identify new seedlings.

As you plant your own native flower garden, you'll learn to observe the intrinsic successions within it, and come to more deeply appreciate the evolved and orchestrated succession garden of the prairie or meadow. And that is just the beginning.

Healthy soils can be measured by the vigor, health, disease resistance, and productivity of the plants you grow.

TENDING THE SOIL

Soil health is the foundation of every garden—and the literal and proverbial roots of our health as a species. The wise and conscientious gardener minimizes soil disturbance in their garden. They feed their soil with compost and leaf mulch, knowing the keys to garden success lie in the complex living ecosystem that is unseen and underfoot. They utilize complete organic fertilizers, mindful of the inputs they gather. This gardener is careful to protect the soil's structure by minimizing compaction. Tending soil means recognizing the vibrant diversity of underground species whose health is paramount to the aboveground foliage and growth of plants.

No-till beds are edged seasonally, and amended with fresh compost annually in fall or spring as we prepare the garden for the next growing season.

OPPOSITE
In-ground raised beds create a harmonious space where the lawn meets the vegetables, and the flowers in the lawn keep pollinators close by.

The Case for No-Till Organic Gardening

We have always made it a high priority to minimize soil disturbance in our gardens. Partly because we like to take the lazy route, and partly because, frankly, it saves time and extends the growing season. No-till gardening (also called no-dig gardening) emphasizes growing without annual tillage, and it's a philosophy gaining ground. An additional benefit is that you can successfully plant your no-till garden with minimal preparation. Instead of making lots of noise and disturbing last year's burgeoning microbial life, you leave your soil intact year to year, trimming plants at the ground level, leaving their roots in place to decompose and nourish those soil microbes.

If you own a tiller, this might seem counterintuitive. Maybe it's even a little bittersweet to imagine the loss of a perennial spring activity that's been integral to your garden life, especially if it's a process that's worked for you. But the time and resources saved by not tilling is enormous, and that alone is reason enough to consider the switch to no-till gardening.

We do not routinely till our garden soil. We did one initial till to work in compost, silt, and sand when we created our beds, an effort to amend our heavy clay

into a beautiful loam. By adding a specific ratio of sand and silt to our dense clay, we shifted the soil structure to one that's more porous and well-draining. In the process, we also created our in-ground raised beds.

Going on year five with a no-till philosophy in our expansive home garden, I can attest this method is a win-win. With fewer inputs, a no-till garden approach is more environmentally friendly. By not disturbing the soil, we bring fewer weed seeds to the surface, preventing them from germinating. Furthermore, we save time by not having to wait for dry spring conditions necessary for tillage before we can plant, thus allowing us to start our growing season sooner. For me, this last part is key: extending our season simply by not tilling is a succession garden win.

Understanding Your Soil Health

Your local agriculture extension is the best place to begin to understand your soil health. For a minimal fee, they will analyze your soil and furnish a report. Ours cost $17, and included soil texture, soil pH, NPK (nitrogen-phosphorus-potassium) amounts, and salinity. Studying your soil is an important first step in planning your garden. Understanding your pH is essential. While there are exceptions, most vegetables prefer slightly acidic to neutral soil (pH 5.5–7.0).

There's a fascinating relationship between soil pH and plant vigor. Availability of nutrients varies by soil pH; not only the macronutrients NPK, but also micronutrients. Phosphorus, for example, is most readily available to plants at a neutral pH, and the soil microbiome, particularly bacteria, thrives with soil pH from 6–6.5. Happy bacteria in your soil will break down plant material and recycle nutrients, making them available to your plants.

Soil pH can be amended if it is too acidic (below 5.5) or too alkaline (above 7.5). The addition of agricultural lime (pulverized limestone) will raise your pH, while adding elemental sulfur will lower it. These are not instant fixes, and take time to make an impact. If your soil pH needs adjusting, it's best to begin researching and amending today.

So much goes on underfoot that we take for granted, but understanding your pH is a great first step in gaining an understanding of your soil, the availability of its nutrients, how you can improve its health, and the productivity of your food garden.

Feeding the Soil

Compost amendments are the cornerstone of our soil's health. While we do make our own compost, it is not enough to supply our entire garden, so we supplement by purchasing compost locally. We are generous with our compost amendments, adding 2 to 3 in. on top of each garden bed annually.

Placing compost on top of the garden beds instead of working it into the soil is strategic. First, it saves a lot of time and effort; that alone is cause enough to make the switch. Second, it acts as a mulch, helping to lock in moisture while also suppressing weeds. We don't mulch with straw or woodchips, only compost. All the while, this cycles nutrients deeper into the beds, helping to build soil structure, feeding the microbes who, in turn, feed our vegetable plants, and eventually us. This shift has been the single best change we've made to our garden philosophy, ever.

Crop Rotation

Planting an area with a different plant family one after another is known as crop rotation. This practice ensures soil nutrients are replenished and disease pressure doesn't get the upper hand in your garden. While it's common in large-scale farming, we have never fully adopted crop rotation in our home garden, in part because we grow so many brassicas annually and don't have enough space to adopt a full crop rotation schedule.

However, we never follow with the same plant family back to back in the same space within the same growing season. We are relentless with annual crop rotation of plant families most fraught with pernicious soil-borne disease, such as tomatoes, as well as any plants that have exhibited fungal disease pressure. Fungal spores can reside for years in the soil (yes, years), surreptitiously lying dormant only to emerge and devastate garden dreams.

Homemade compost is a new joy for us, brought with space to make massive annual heaps. This is our 2017 compost pile, and we plan to ceremonially spread it in 2022 in our espalier orchard.

With space at a premium even in our large, rambling garden, we instead lean heavily on tending our soil with good nutrition to compensate for the lack of a full crop rotation schedule. The most extensive crop rotation plan I've seen is a seven-year rotation, and that's too much to keep track of for me. We shortcut this lengthy schedule by supplementing annually with balanced, slow-release organic fertilizer that replenishes macro- and micronutrients consumed by preceding crops, keeping the soil healthy and our plants happy.

Slow-Release Organic Fertilizer

We feed our soil with organic fertilizer in addition to compost. We use it as a broadcast fertilizer over entire beds, and mix it directly into transplanting holes. We use it everywhere we grow food, across our entire orchard and throughout our vegetable beds, in similar proportions. I love the simplicity of this "one size fits all" approach, and the gentle but effective nutrition it provides our plants. Because it is slow release, I can be confident our plants will have the resources they require at their disposal for several months.

A handful (roughly 3–4 tbsp. per scoop) of slow-release organic fertilizer is mixed into each planting hole for each transplant in our garden, even our flowers.

Making Organic Fertilizer

We found the original recipe we used for organic fertilizer in one of our first gardening books, *Growing Vegetables West of the Cascades*. At that time, we lived and gardened in that area, so it was the perfect accompaniment to our early gardening career. What we didn't realize is that the heavy clay prevalent in western Oregon is also quite common in the Midwest, and this book has remained an invaluable asset throughout the decades. I have been fortunate to connect with the author, Steve Solomon, via email, and he has graciously shared his updated complete organic fertilizer recipe with me. Lovingly named "Solomon's Gold," it's a tried and true recipe he mixes at his home in Tasmania today. It's an essential organic gardening supplement, and I've modified it slightly to include more easily procured materials.

COMPLETE ORGANIC FERTILIZER

Nitrogen
- 50 pounds (23 kg) soybean meal or other seed meal

Calcium
- 22 pounds (10 kg) agricultural lime
- 10 pounds (4½ kg) agricultural gypsum

Phosphorus
- 15 pounds (7 kg) rock phosphate

Potassium
- 2 pounds (1 kg) potassium sulfate

Trace elements and micronutrients
- 6 pounds (3 kg) kelp meal

Optional ingredients as determined by soil test
- 2¾ ounces (78 g) copper (copper sulfate)
- 2¾ ounces zinc (78 g) (zinc sulfate)
- 1½ ounces (42½ g) boron (Solubor)
- 7 ounces (198 g) manganese (manganese sulfate)
- 1½ ounces (42 ½ g) molybdenum (sodium molybdate)

Mix all ingredients in a wheelbarrow or large container. We like to add a proportional amount of each ingredient, mix well with a shovel, and repeat until it has all been incorporated. We always mix it up again before we pull from it. Store in a covered container.

Cover Crops

Buckwheat is a very fast cover crop that will happily self-seed, so right after the flowers have finished, cut the plants down, leaving roots intact to further benefit the soil.

Cover crops are a clever early-season placeholder for your late season garden dreams. If you're like me, open space is a tantalizing invitation to fill it now, right away, capitalizing on every square inch, with no line of sight or care of the seasons ahead. Immediacy drives these decisions, and with plants in hand, it's hard to argue with what today provides. To combat this, about a week or two after our last frost, we seed buckwheat over the square footage dedicated to those distinctly cooler (and, at that point, distant) months. This practice consistently curbs my enthusiasm and provides a personal sense of accomplishment. Meanwhile, the garden benefits in multiple ways.

Buckwheat is a fast-growing crop good for both soil and pollinators. Flowering in mere weeks and maturing in less than two months, it's a friend to the short growing season gardener. Reaching heights of 2 to 3 ft., it quickly establishes and blooms profusely, smothering weeds and retaining soil moisture. It self-seeds abundantly, so after it's flowered and before it sets seed, we simply cut the planting at the ground level and compost the phosphorus-rich plant material. Soon after that, we transplant the slowest growers of our main season into the bed: fall cabbages, broccoli, and cauliflower.

Soil health is the foundation of our health as a species. It is imperative for a healthy garden and a healthy planet. Taking the time to understand and amend your garden soil is the key to healthy plants. Healthy plants withstand disease and pest pressure more readily than plants that are stressed. Just like with our bodies, what we feed our soil is what fuels our plants, and thus their immune systems and ability to combat disease. Feed your plants well and they will return the favor in plentitude.

A succession garden is a continuous evolution, negotiating space and time toward the goal of consistent harvests spread out across as many seasons as possible.

SEED STARTING AND GARDEN PLANNING

Succession gardening triumphs when it is implemented across spring, summer, fall, and winter. Each comes with its own unique opportunities and tasks that have the potential to boost your garden's productivity. Succession gardening embraces seasonal eating, and relishes the flavors each season has to offer. It capitalizes on every opportunity to lengthen the growing season. As you will see, it's truly never too late (or too early) to begin.

My entire frame of reference for the seasons shifted when I embraced the notion of continuous planting for continuous harvesting. That sense of foreboding and urgency dwindled, replaced by an open invitation to renew the garden, and often. You can transform a completely disease-riddled bed in the height of summer and renew it, cutting short one succession in favor of a more disease-resistant, more quickly maturing or cold-hardy option. You can delay plantings to align with your stamina for processing hefty harvests the likes of pickling cucumbers and paste tomatoes. By preparing garden beds in fall, you make ready to sow and transplant as soon as the spring thaw arrives. Above all, planting diversity across the four seasons cultivates enjoyment and guarantees productivity.

Connecting your individual rhythm to the seasons is a highly personal endeavor, best shaped by your own goals for your garden. While establishing perennials is best prioritized in the early years, annuals are a boundless playground of diversity that promise an infinite source of exploration. Gardening in the shoulder seasons shines with plants like leafy greens, root vegetables, and a deep bench of brassicas, all of which thrive in cooler weather.

Fully planted, or ready for new successions? The garden is never static except in photos; plants like this bok choy mature, and new seeds are sown or transplants tucked in, almost weekly for nearly six months straight.

GARDEN SPACE AND HARVESTS BY SEASON

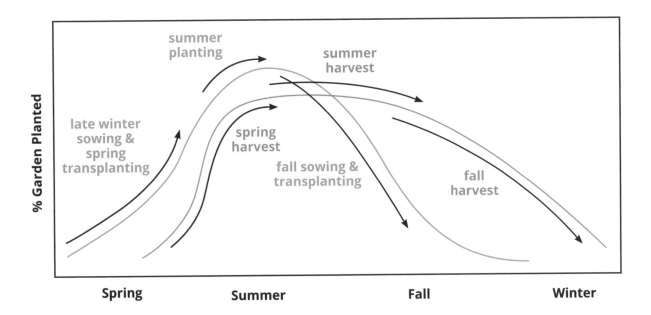

For us, overlapping sowing and harvesting schedules climax in the first part of summer. After that, the garden shifts more toward harvesting, though we continue to sow until very late summer in our cold climate—in warmer zones, you can sow for even longer. Establishing and tweaking your seasonal tasks will help you to develop best practices and routines particular to you and your garden.

As with any great endeavor, garden success hinges on a strong plan that is well executed. In an ideal world, you could imagine every possible scenario and weigh them all, creating and recreating your garden plans until the perfect one settles clearly in your mind's eye, and you execute it with confidence. Reality narrows the focus to something in between.

Bringing garden plans to life in a short and cold climate both invites and necessitates creativity. Time is my biggest threat to succeeding as a succession gardener. Start too early, and my plants will suffer from shock and stunted growth, challenges from which they may never fully recover. Start too late, and they may miss their window of opportunity to thrive, and either succumb to pest pressure or skip right to seed production.

Winter's Wisdom

Winter in a northern climate is nothing short of remarkable—and difficult, frigid, and generally a month or more longer than mental fortitude can gracefully weather. But when you pause and appreciate all the life just outside your cozy home, dormant in utter stillness, it is a miraculous feat. Plant and animal life lay in wait, alive yet frozen solid. Fruit trees stand tall, enduring all that winter thrusts at them, and go on to become some of the first flowers of spring. Prairie seeds lean into this extended cold and wet stratification period to prepare for spring germination. Gardeners who follow their lead get ahead in the garden too.

OPPOSITE
Dormant prairie standing tall all winter long, providing forage for wildlife, and slowly preparing its seeds for spring's germination frenzy.

ABOVE
With our earliest brassicas established in the low tunnels, it's time to interplant these beet and head lettuce starts, which will mature well before the cabbage and broccoli.

ABOVE RIGHT
Winter seed starting is a marathon, not a sprint. It starts slowly, but by April, our deck is a veritable garden center most days as I shuffle plants inside and out for some full-strength sunlight.

The growing season starts out at a pace commensurate with the stillness that surrounds us during the shortest days of the year. These are the weeks and months of tending slow-growing vegetable and flower starts, those that require a head start indoors to comfortably succeed in our short growing season. We sprout organic seed ginger in late January, making tropical dreams a reality even at 45°N. By the end of February, August's tomato, pepper, and eggplant dreams have been seeded, along with the hope of a large bin of storage onions. More seeds are sown every week. A fluid-yet-detailed procession of seed starting helps stretch how long we grow.

There's a delicate balance between starting seeds early enough but not too soon. It requires significant resources to maintain vigor and health of seedlings indoors before outdoor conditions are suitable for transplanting. If you're a zone pusher and enjoy rolling the dice in the early season, this includes taking calculated gambles to ensure you produce large, vigorous starts by your earliest possible transplant date. It also means you are prepared to lose some transplants to inclement weather, and have backup starts just in case. Balancing the goals you set with your available resources will help you determine which plants should get your undivided attention and most focused resources, and which can be purchased as starts or sown directly in the garden when conditions are ideal.

WINTER TASKS

Seed starting	Assemble light setup before you need it
	Check all lights; damp dust to improve light emissions
	Replace older bulbs to maintain full strength and spectrum
	Clean seed-starting trays in preparation for sowing
	Mix seed-starting soil mixture
	Start with slowest growing plants: asparagus and alliums first, then peppers, eggplants, and tomatoes beginning late February through April
	Push early season with super early flat of brassicas, indoor sown ten weeks before last frost; transplant six weeks before last frost (zone 6 and cooler need extra protection to transplant this early)
Low tunnels	In late February, add plastic sheeting to low tunnel(s) to begin melting any snow and thawing soil; winter sun is crucial for this phase of season extending
	In mid- to late March, sow first radish, carrot, pea, and arugula seeds; transplant earliest cold-hardy vegetables
Perennials	In late winter, prune fruit trees and shrubs
Potting up	Plan to pot up slow growing crops as needed; we prefer compostable newspaper pots in both 3¼-in. and 4¼-in. size

The Earliest Succession

The key to being prepared for early spring in the garden is a dependable and trusted seed-starting system that produces strong plant starts. The alternative is to purchase starts for slow-growing vegetables—including onion, tomato, pepper, eggplant, and woody herbs such as rosemary and sage—and to direct sow everything else. Investment in your seed-starting system rewards you with more than just healthy starts, it provides a wellspring of hope as frigid winter days turn to weeks and months. During these prolonged weeks of fluffy white precipitation, it is far more joyful to measure the growth of our seedlings than the accumulation of snow on garden beds.

The Merits of Growing from Seed

Changing your relationship with seeds is one of the most meaningful steps you can take in extending your gardening season. Our penchant for seed starting has ballooned right along with our garden, growing at a rate that matches its footprint. Since our earliest days and first container tomatoes, selecting, ordering, and growing from seed has always been the norm, and buying starts the exception, save for bedding flowers. As the years passed, we diversified our vegetable grow list, and we grew right along with our plants. This progression of knowledge was naturally and intentionally slow, driven strictly by our hearts and palates. We started small and made measurable, incremental changes with each new season.

You must capture this earliest season. Growing from seed, as opposed to shopping for starts, exponentially expands your potential diversity from, say, the few kale varieties at the garden center, to dozens of shapes, colors, and textures. More importantly, it puts you in full control of when your growing season begins and with what. Growing from seed affords you the space and time to deeply reflect on your goals, and seek out seeds that match those edible dreams, visually and culinarily. Seed starting is also the fuel that boosts our growing season well beyond what is readily available, providing a personal nursery of vegetable seedlings from March through August, and homegrown food from late April to well beyond the New Year.

Our indoor seed-starting system is pivotal to our succession garden. The process includes starting enough for unforeseen challenges, which, when not needed, become donations to friends and community gardens.

OPPOSITE
Hardening off seedlings for our early low tunnels means exposing these cold season champions to a sunny, mild 38°F (3°C) day after a snowstorm.

I have yet to visit a greenhouse in late March in Minnesota selling starts of stout, hardened off kohlrabi, bok choy, cabbage, broccoli, or even kale, precisely the types of plants I need at that moment for the extra early transplanting that extends my growing season. I imagine they're tending their first flats of young brassicas around the time I'm transplanting my hardened-off starts under low tunnels. Consumer demand for those plants isn't high until early May here, the typical start of the gardening season.

A bit of planning, record keeping, natural curiosity, and a desire to continuously improve is all the fuel you need to add that extra succession to your garden. And you'll gain valuable square footage in the height of summer after you harvest these earliest crops—and then plant a beautiful fall garden in succession. This is the magic of season extending in spring.

Indoor Seed Starting as a Way of Life

We center our lives around the garden, so it should come as no surprise that our grow lights and plant stands brighten our main living space come late winter. Positioned prominently in our heated living area rather than our chilly basement, the ambient air and soil temperatures there create ideal growing conditions. Our starts' position in the sunny, south-facing bank of windows also makes for a constant reminder to tend those little pots. We naturally gravitate to them at least a few times a day, checking soil for dryness or signs they've sprouted, so we can remove the germination dome. Self-watering trays are the single biggest improvement we've made to our seed-starting system over the years; they provide even and consistent moisture.

SEED-STARTING SUPPLY CHECKLIST

- Seed
- Adjustable lights and plant stand
- Power strip, extension cord, and timer
- Germination mats and clear domes
- Self-watering trays *or* 2-in. soil blocks, soil block trays, base tray, and wicking mat
- Well-balanced seed-starting or soil block mixture
- Small fan (to strengthen stem turgidity)

Etiolation, the scientific term for stretched stems, occurred with these larkspur seedlings, who require darkness to germinate. I salvaged some, but luckily self-seeded larkspur in the garden proved stronger and germinated quite early.

OPPOSITE
We have traditionally grown under T8 fluorescent lights, four-bulb ballasts each housing two trays of seedlings. We leave the seedling mats on for a minimum of one to two weeks post-germination to accelerate establishment.

Input Drives Output

Sunlight sustains life, and implicit in successful seed starting is a robust system with sufficient light, natural or artificial, to encourage sturdy stems and lush foliage. It sounds easy enough, and of course we all have good intentions, but this requirement proves challenging time and again, even for the seasoned gardener. If you're committed to growing from seed and will be sowing indoors, this singular act is the foundation of your garden's vigor and success, so it's worth taking the time to get it right.

When plants lack sufficient light, they quickly respond, even to the sunniest, south-facing windowsill or artificial light. Plants adjust through *etiolation*, or stretching themselves, compromising stem turgidity in favor of photosynthetic capacity, in hopes of becoming stronger for it. More often than not, however, this results in irreversible damage, their stretched stems then unable to withstand the realities of outdoor growing conditions. Some seedlings can outgrow this condition through deep planting, but your best course of action is prevention, positioning lights as low as necessary to encourage strong, stout plants.

Double Duty: Less is More

This soil block legend is a complete succession planting guide. It contains the date I sowed the tray and the specific variety of each vegetable, plus the amount of each variety I sowed and plan to plant in the garden. For example, Tiara is an extra early, compact, hybrid green cabbage, and Blue Wind a compact, early season broccoli. Beas, Kolibri, and Terek are three varieties of kohlrabi we enjoy growing. This legend is my road map for checking germination rates and planting out my trays. Kept online, this legend is accessible from my smart phone, so it's portable and always available for reference. Adjustments made are noted in real time.

These spreadsheets are the foundation of our seed-starting routine, our preferred method for keeping track of sowing dates, varieties, and quantities sown. The vegetables in this tray will mature from late April (bok choy) all the way into November (kale), with most maturing between early May and late June. A single tray includes multiple successions of food that will mature across the growing season.

I often reference these spreadsheets to track how many weeks or months varieties take to mature at various times of the growing season. Every time I sow another tray, I add another legend to the sheet. Every time I direct sow, I record the date and linear footage of what I sowed, also noting variety, so we can accurately track the total square footage of carrots sown, for example, and how they perform for us. The goal is that one year, our sowing charts will be largely dialed in, and all that may change are varieties as seed sources fluctuate.

EXAMPLE SEED-STARTING SPREADSHEET

February 28									
Tiara	Romanesco	Bluewind	Famosa	Integro	Joi Choi	Scarlet	Beas	Kolibri	Terek
Tiara	Romanesco	Bluewind	Famosa	Integro	Joi Choi	Scarlet	Beas	Kolibri	Terek
Tiara	Romanesco	Bluewind	Famosa	Integro	Joi Choi	Scarlet	Beas	Kolibri	Terek
Tiara	Romanesco	Bluewind	Famosa	Integro	Joi Choi	Scarlet	Beas	Kolibri	Terek
Tiara	Romanesco	Bluewind	Famosa	Integro	Joi Choi	Scarlet	Beas	Kolibri	Terek

The Right Stuff

Our trusted and time-tested seed-starting system includes a 4 ft., four-bulb fixture for each shelf. Germination mats aid in that process by keeping the soil warm, and 2-in. soil blocks, complete with self-watering trays, provide even moisture. Again, seedlings started indoors need strong enough light to develop sturdy stems that will stand up to the next pre-garden phase, hardening off. The investment in lights and plant stands were their own successions in our garden journey: we added one plant stand each year until we achieved ample indoor space commensurate with our outdoor space.

Positioning your plants close to the light will encourage lush growth without legginess. We aim to keep seedlings within an inch of the fluorescent lights for the duration of their time indoors. We fit two standard trays (100 seedlings) under a single 4 ft. ballast light. I have witnessed firsthand how quickly plants become leggy, even under artificial light, if not close enough to the faux sun. Once they've stretched, there is little relief to this physiological damage.

Our indoor growing space is some of the most prized real estate in our succession garden. It helps add as many plantings as possible to our all-too-short growing season. It's one of the reasons our bell peppers are blocky and ripe by early August in our chilly zone 4 garden. It is how I keep brassica and lettuce starts ready throughout the growing season.

Starting seeds indoors without supplemental lighting, even in your sunniest, south-facing window, will not yield favorable results. I do not recommend indoor seed starting without artificial light—fluorescent or LED. There simply isn't strong enough light coming through the window. If this is your situation, direct seeding your garden is your best option, because that will actually produce stronger plants, but be aware harvests from a direct-seeded garden will be delayed upwards of six to eight weeks compared to an early spring garden with transplants and frost protection.

The LED vs. Fluorescent Experiment

It's been 20-some years of growing vegetable starts in winter. Some years they were on our kitchen floor, others in a chilly basement, and more recently, they've nearly taken over an entire sunny room in our home. We have always used fluorescent bulbs. We started that way, and it works, so it's hard to reason with that logic, especially because it means potentially throwing away well-functioning ballasts.

Recently, however, as the lights started to lose their strength and warranted replacement, we decided to experiment with LED equivalents in our existing four-bulb fluorescent fixtures.

A few modifications were made to house the bulbs, and identical trays were sowed. What we hoped to observe was near-identical growth from each tray, which would definitively give us the go-ahead to begin phasing out fluorescents, and phasing in more energy-efficient LEDs. So far, replacement bulbs at a cool white 6,400 Kelvin seem to produce seedlings of similar quality. The biggest change is that the LED lights are positioned much higher from the plants than fluorescent equivalents. All in all, we're pleased with the results, and will continue to transition our lights to LEDs as the fluorescents need replacing.

Soil Blocking: A Pot-less Seed-Starting Alternative

Soil blocks are a wonderful alternative to plastic seed trays. They require an initial investment in trays and a blocker, but free you from those flimsy plastic plug trays. Their merits include a self-contained pot for the seedling as it grows and its roots spread and help build structure. They are, in essence, a pot-less pot.

Soil blocks are a porous medium with moderate nutrition. They are available in four sizes: ¾ in., 1½ in., 2 in., and 4 in. We invested in just one size, the 2-in. soil blocker. It's large enough that we don't fret about constantly watering, and spacious enough that a seedling can happily live out its first four weeks before transplant without running out of room.

The concept of soil blocks was brought across the pond from the United Kingdom by Eliot Coleman and popularized in his garden tome, *The New Organic Grower*. We have tried varying recipes by substituting coconut coir for peat moss, but these trials have been met with mixed results. I wouldn't recommend a complete substitute of coir for peat, but half and half would be a good compromise.

SOIL BLOCKS

- 3 parts peat moss
- 1 ounce (2 tablespoons or 28 g) lime
- 2 parts compost
- 2 parts perlite or coarse sand
- 1 part garden soil

We use a half-gallon (2 qt. or 2 kg) yogurt container for our "part." This recipe yields about 150 2-in. soil blocks. First, we mix the peat moss with the 1 oz. (2 tbsp. or 28 g) of lime to increase the pH of the acidic moss. (Note that lime, the only ingredient measured differently in this recipe, is proportionate to the rest, so adjust accordingly) Then we mix in the perlite, which we immediately dampen because of its dustiness. Finally, we add in the compost and garden soil, mixing thoroughly and wetting well.

Let this mixture sit for an hour to fully absorb the water, and check for wetness. You should be able to squeeze water out of a handful, and the mixture should be quite wet, but not full of standing water. The amount of additional water you add will be determined by how much moisture was in your bulk materials when you initially mixed them. It is easier to add more water than wait a day for the mixture to reach the right consistency due to over-wetting. I speak from experience.

What too soon looks like: one gallon tomato pots in early April, a good three weeks before the garden can accommodate them under frost protection. Here, they're waiting in limbo for warmer garden days.

OPPOSITE
A perma-nest tray with PVC blocks cut to length creates space for a water reservoir. Nylon woven through the soil block propagation tray and a capillary mat work together to water the soil blocks from below.

Investing in indoor growing space in colder zones is perhaps the biggest pay-off for the succession gardener. And while my family has expanded our space, it continues to prove challenging to meet the increasing demands of our hopes and dreams come early spring. Seedlings swiftly stretch across all four plant stands in two areas of our home. The indoor seed-starting garden is now sprawling in proportion to our outdoor garden, and when faced with prioritizing limited space, innovative choices must be made.

While beets and lettuce can be sown indoors to get a jumpstart, if you only have one light, focus it on the longer-maturing crops like cabbage, broccoli, onion, tomato, pepper, and eggplant. These need longer head starts indoors before they're garden-ready, and won't do well direct seeded. We generally start this diverse group of garden suspects in our home by eight to ten weeks before last frost. While the cabbage and other brassicas head to the garden before April 1, the rest remain indoors, potted up, under lights, and as weather permits, we slowly begin the hardening off process.

Unless you have the indoor space and deep desire to extend your growing season, resist the urge to start your plants too early. If you are not planning to season extend with hoops, wait to sow the majority of your seeds until six weeks before your last frost—the exception being if you have sufficient indoor space with adequate light; are excited about potting plants up; will enthusiastically commit to the daily plant shuffle while hardening off; and love the idea of coddling seedlings who will become unruly teenagers.

WINTER SUCCESSION PLANTING GUIDE FOR EARLY HARVESTS

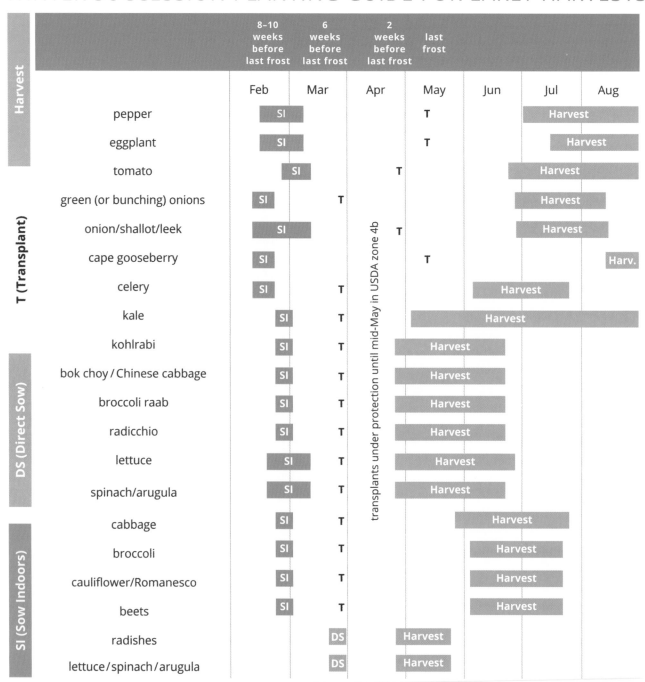

Harvest		8–10 weeks before last frost	6 weeks before last frost	2 weeks before last frost	last frost				
		Feb	Mar	Apr	May	Jun	Jul	Aug
T (Transplant)	pepper	SI			T		Harvest	
	eggplant	SI			T		Harvest	
	tomato		SI		T		Harvest	
	green (or bunching) onions	SI	T				Harvest	
	onion/shallot/leek	SI			T		Harvest	
	cape gooseberry	SI			T			Harv.
	celery	SI	T			Harvest	
	kale	SI	T		Harvest		
DS (Direct Sow)	kohlrabi	SI	T		Harvest		
	bok choy / Chinese cabbage	SI	T		Harvest		
	broccoli raab	SI	T		Harvest		
	radicchio	SI	T		Harvest		
	lettuce	SI	T		Harvest		
	spinach/arugula	SI	T		Harvest		
SI (Sow Indoors)	cabbage	SI	T			Harvest	
	broccoli	SI	T			Harvest	
	cauliflower/Romanesco	SI	T			Harvest	
	beets	SI	T			Harvest	
	radishes		DS		Harvest		
	lettuce / spinach / arugula		DS		Harvest		

transplants under protection until mid-May in USDA zone 4b

Winter seed starting and succession planning includes sowing vegetables from each succession type: quick, midseason, and late season. All started around the same time in late winter, yet their harvests spread out from late April until well after first fall frost. These earliest successions are the most diverse, and provide the greatest opportunity to spread your harvest across the longest time horizon.

The Art of Hardening Off

Plants that are raised indoors and even in greenhouse settings are pampered. They are protected from wind, the very thing that teaches them strength and develops stem turgidity. They are sheltered from the sun, supplemented with artificial light instead, their tender foliage not as sturdy as plants grown outdoors with full strength sunlight. And they grow at ideal temperatures. The transition from these spa-like conditions inside to the real world can be harsh—and even fatal—unless you take the time to introduce your plant babies to the real world. Change is hard, even for plants.

Hardening off is a bit of an art—and a test of patience. It's basically garden purgatory. It takes at least one full week to do well, your plants receiving a little more sunlight with each day until their leaf structure has quite literally hardened off, forming thicker cell walls. As you start with part shade and work your way to full sun, time stands still between the excitement of seed starting and the joy of actually gardening. If you've been a good plant parent, keep at it, and take your time with this phase. Don't expose your starts to a full day of sun right away. Ease them into it.

We have the benefit of dappled spring sunlight and summer shade on our deck, and use this space for the first phase of hardening off. It also happens to be the closest outdoor space to our growing lights, so whether or not the light is ideal is rather inconsequential. I tend to go with the easy route, and this happens to be the easiest path to getting the plant babies physically outdoors. Even with lingering snow, I usually begin the hardening-off phase for my earliest cold-hardy brassicas during the third week of March. Cloudy days, preferably above 40°F (4°C) are also ideal at first, because I know I can leave plants out longer without suffering damage—which, honestly, happens more than I'd like to admit.

Slow and steady with the flower starts, utilizing partial shade from dormant tree branches and being inventive as needed to give these plants a soft landing in the great outdoors.

Only when my starts have spent part of their days outdoors in partial to full sun will I roll the dice and leave them out longer. An evening low of 40 to 45°F (4 to 7°C) is fair game for these cold-hardy seedlings to have a sleepover on the deck, although at first, I usually tuck them in under a frost blanket to retain a little warmth. By the end of that first week, any cloudy day will be a full day spent outside. Still benefitting from part shade during the day, they'll be well on their way to being ready for the garden. The second week, I leave them outside all day long, assuming air temperatures are above 45°F (7°C) and the wind isn't too stiff and chilling.

Hastening the hardening-off process will result in literal plant sunburn, and can be quite the setback at this tender phase.

Once our earliest transplants are hardened off, they end up going into a protected environment (a low tunnel), so we're largely hardening them off for intense sunlight and air temperatures. When it's feasible, I transplant right before a stretch of cloudy days. An imminent snowstorm is a perfect cloudy day too, for this adventurous, early spring succession. For warm-season vegetables such as tomatoes, whose time is several weeks to a month later, I'd hope for a little cloudy weather also, but quickly followed by warm, sunny days. Just ahead of rain is my next favorite weather pattern for transplanting or seeding the garden—and weather always dictates my gardening schedule.

You may be wondering, "If you can direct sow most crops, why go through the extraneous process of seed starting indoors?" With only four and a half months of frost-free growing days, we are in a race against time in our growing zone. If I were to direct sow the same early season garden, I would lose four to six weeks of opportunity at the height of summer waiting for the first direct-sowed succession to mature. By sowing indoors instead of using garden space, I am growing the equivalent of a direct seeded garden in late February, a feat only possible at least two zones south of me. Zone bending, as I call it, ensures the majority of the garden is available for planting with hot season crops for the longest possible stretch during the hottest weeks of the year. This is how indoor sowing adds significant productivity to our garden.

Sowing Hope

Getting ahead without all the fuss is easily attainable too. When you time your earliest sowings six weeks before your last frost, you will be met with an easier, albeit later, start to your garden's season. During those warming days, you will harden your plants off leading up to your last spring frost, and perhaps won't even need to bring them in every night. It's much easier logistically, and will minimize any potential legginess or pot-bound misfortunes lurking in the shadows of lengthier indoor stints. Your garden will be ready to receive the starts when they're properly hardened off, right around your last frost date. This is undoubtedly the simplest way to grow from seed for the home garden.

Lean into early season vegetables like radish, spinach, peas, and arugula, sowing them when soil temperatures are between 40 and 45°F (4 and 7°C). This is the simplest, most no-fuss approach to getting ahead in spring.

Do lettuce and beets grow well directly seeded in the ground? Absolutely. The limiting factors to this approach in the early season are your soil and air temperatures, which may slow germination and establishment. If direct sowing is your preferred method, then your early season garden will germinate and yield a little later, as you're at the mercy of soil temperatures to break dormancy. You could also supplement with plant starts as soon as you can find them at the local nursery. And the rest of your garden will be implemented a little later, thus pushing the entire stream of harvests further into summer. In cold climates, I would argue those are reasons enough to hone your indoor sowing skills and hasten your spring harvest. In some climates, the difference between one or two extra weeks of growth is dramatic. It could mean the difference between vine- or countertop-ripened homegrown tomatoes, because an early frost descended.

While indoor seed starting is our biggest ally and deepest joy in our succession garden, direct sowing is a method we welcome too. While the seedlings sown under row cover emerge sooner and establish faster than those planted in uncovered beds, the uncovered radishes, for example, are not far behind by late April when harvests commence under cover. Either way you sow, those foods are harvested before or around last frost—and even without all the fuss of low tunnels and indoor sowing, I guarantee there are ways you can get ahead in early spring quite effortlessly and extend your growing season.

Seed starting and the planning that goes along with it are what lead the way in the resilience of our succession garden. Today's garden must be equipped to meet the more customary seasonal challenges, as well as the perils of climate change: more extreme weather patterns, late spring frosts, flooding, drought, high winds, hail, and early fall frosts. These strategies ensure your garden will be flexible and adaptable, enabling you to make the most of every situation.

By mid-April, sowing trays are a motley crew, including tomatoes, peppers, onions, and our early major harvests, the beloved brassicas.

SPRING AHEAD: HASTENING THE GROWING SEASON

Spring is the season when hope and renewal return. As a succession gardener, this is the time when you visualize your entire growing season. It's the season of planting both quick successions that will mature in weeks and late season successions that will anchor the garden through autumn. It is the time of steadily sowing and transplanting vegetable varieties with varying succession rates to maintain a steady flow of food for the coming months. It's an intense season of planning and execution. Any task left for another day becomes a task for that moment. This makes for long yet rewarding days and sleepy nights.

My most treasured garden activity is taking our blank slate and recreating it every spring. The garden is slowly planted across the weeks of late winter through late spring, tailored to meet our needs and designed to spread out the harvests. By late March, I have a budding group of vegetables and flowers inside under lights and outside under row cover. The foundation of our succession garden is sowing the same brassicas a few times in spring, and planting varieties with a wide range of maturation dates and tolerance for cold and heat. I build on this foundation weekly during this nascent season of opportunity.

Sowing Early and the Case for Low Tunnels

A key tenet of succession gardening is season extension. It's easiest to execute during those yearning weeks of late winter, and a simple DIY low tunnel creates a microclimate one or two zones warmer. A commitment of your resources is needed to achieve these extra successions. What you gain for your investment and effort is more self-sufficiency, a sense of deep pride, and in case that's not enough, probably the earliest vegetables around.

Our earliest cool season transplants are tucked into the garden in late March under the warmth and protection of row cover. In our climate, this is about six weeks before our last frost. Our first year implementing this strategy was met with a few restless nights, especially when the lows were well below freezing and snow accumulating by the hour. To my delight, the plants more than survived, and went on to produce our earliest brassicas ever, further fueling our curiosity. Time and again, this has proven to be a feasible and successful way to add another succession to our garden. Now an annual tradition, our strong, cold-hardy starts and a little extra care reliably yield food earlier than we ever imagined possible.

As I've mentioned, extending your season is a strategy I highly encourage you to explore. While it's not necessary in warmer climates, it significantly increases the growing season in colder climates. Low tunnels make the most of the shoulder seasons, those extra weeks you didn't realize were actually part of your growing season. And if not for the sheer benefit of the food produced, the mental and emotional benefits of getting outside during winter and working

toward the distant promise of warmer days is often the most precious harvest during these season-extending ventures.

I could not imagine waiting until May to start gardening outdoors, though I recognize this is neither a goal nor a desire for all gardeners. Season extending requires extra resources, and proactively working in less-than-ideal conditions, like religiously shaking heavy April snow off a fully planted low tunnel. Without these measures, you can easily get ahead by planning for cold-hardy crops to be planted out, after hardening them off, one to two weeks before your last frost. You can explore how much earlier your growing season can commence without anything extra, just by observing and trialing different cold-hardy vegetables ahead of your last frost. I guarantee your growing season will expand in new ways.

Middle of April, and the brassicas have established and are well on their way—and we are still four weeks before our last frost.

While we are harvesting our first meager baskets of food, mostly greens and radishes and maybe an early kohlrabi, the majority of home gardens have yet to be planted, both in our region and in warmer climates alike. These extra early transplants would not thrive without proper soil temperatures, and we achieve that by setting up the low tunnels four weeks before transplanting. As the sun starts to return north, its warmth is trapped inside these structures, and warms the soil to 50°F (10°C) before it's time to transplant. We gain four to six weeks of spring gardening thanks to the low tunnels.

The First Transplants

As soon as soils read 50°F (10°C) at 6 in. deep, and the long-term forecast doesn't project nighttime lows below 20°F (–7°C), we transplant our earliest starts. It's a ceremonious day, and hopefully a cloudy one, usually before March 31. Buoyed by increased solar gains each passing day, the cold-hardy transplants adjust quickly. Within a few weeks of planting, they amaze me with their exponential growth, despite (or perhaps in spite of) winter's ongoing battering just outside.

A digital thermometer is a friend to the season-pusher. I start testing soil temperatures around the second week of March, impatiently watching the digital mercury rise to the magic numbers: 45°F (7°C) for direct sowing and 50°F (10°C) for transplanting.

GARDEN SPACE AND HARVEST BY SEASON

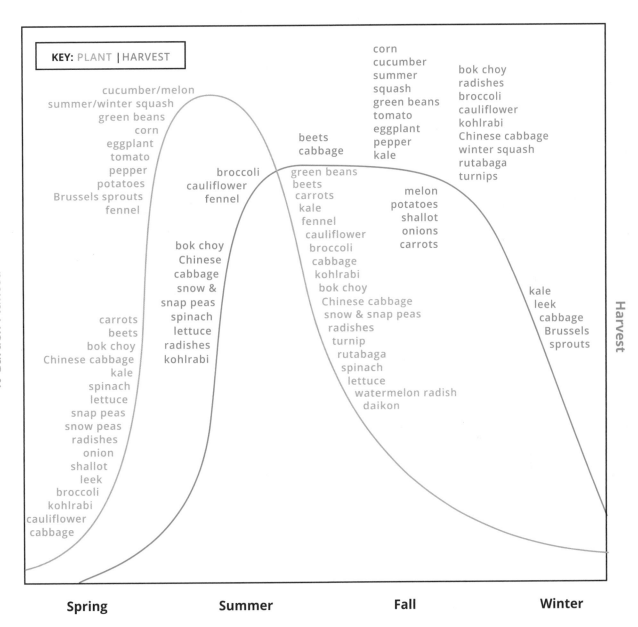

KEY: PLANT | HARVEST

% Garden Planted

Harvest

cucumber/melon
summer/winter squash
green beans
corn
eggplant
tomato
pepper
potatoes
Brussels sprouts
fennel

broccoli
cauliflower
fennel

bok choy
Chinese
cabbage
snow &
snap peas
spinach
lettuce
radishes
kohlrabi

carrots
beets
bok choy
Chinese cabbage
kale
spinach
lettuce
snap peas
snow peas
radishes
onion
shallot
leek
broccoli
kohlrabi
cauliflower
cabbage

beets
cabbage

green beans
beets
carrots
kale
fennel
cauliflower
broccoli
cabbage
kohlrabi
bok choy
Chinese cabbage
snow & snap peas
radishes
turnip
rutabaga
spinach
lettuce
watermelon radish
daikon

corn
cucumber
summer
squash
green beans
tomato
eggplant
pepper
kale

bok choy
radishes
broccoli
cauliflower
kohlrabi
Chinese cabbage
winter squash
rutabaga
turnips

melon
potatoes
shallot
onions
carrots

kale
leek
cabbage
Brussels
sprouts

Spring   Summer   Fall   Winter

Can you see how there's always something to plant in the succession garden? Step into this flow chart at any point in the season and you can capture an extra succession or two.

SPRING SUCCESSION PLANTING GUIDE
FOR CONTINUOUS HARVESTS

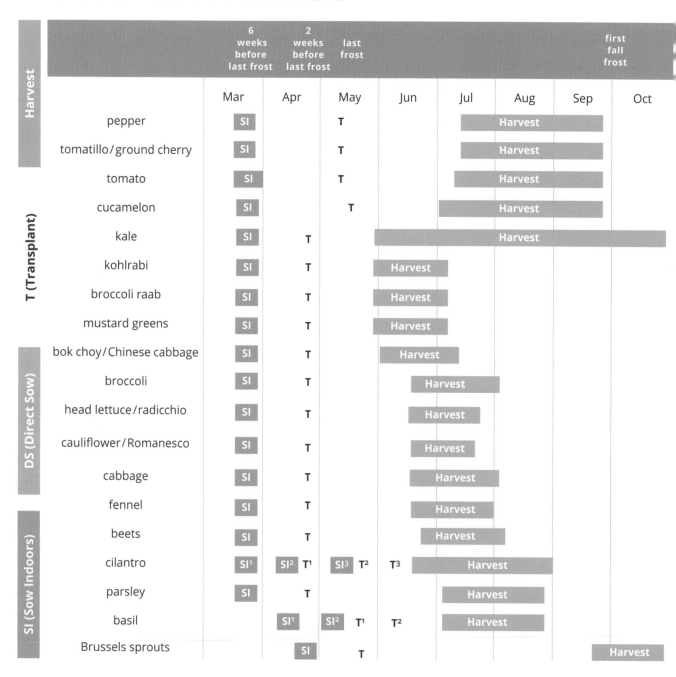

Harvest		6 weeks before last frost	2 weeks before last frost	last frost						first fall frost
		Mar	Apr	May	Jun	Jul	Aug	Sep	Oct	
T (Transplant)	pepper	SI		T		Harvest	Harvest	Harvest		
	tomatillo/ground cherry	SI		T		Harvest	Harvest	Harvest		
	tomato	SI		T		Harvest	Harvest	Harvest		
	cucamelon	SI		T		Harvest	Harvest	Harvest		
	kale	SI	T		Harvest	Harvest	Harvest	Harvest	Harvest	
	kohlrabi	SI	T	Harvest						
	broccoli raab	SI	T	Harvest						
	mustard greens	SI	T	Harvest						
DS (Direct Sow)	bok choy/Chinese cabbage	SI	T	Harvest						
	broccoli	SI	T		Harvest					
	head lettuce/radicchio	SI	T		Harvest					
	cauliflower/Romanesco	SI	T		Harvest					
	cabbage	SI	T		Harvest	Harvest				
	fennel	SI	T		Harvest	Harvest				
	beets	SI	T		Harvest	Harvest				
SI (Sow Indoors)	cilantro	SI[1]	SI[2] T[1]	SI[3] T[2]	T[3]	Harvest	Harvest	Harvest		
	parsley	SI	T			Harvest	Harvest			
	basil		SI[1]	SI[2] T[1]	T[2]	Harvest	Harvest			
	Brussels sprouts		SI	T					Harvest	Harvest

184

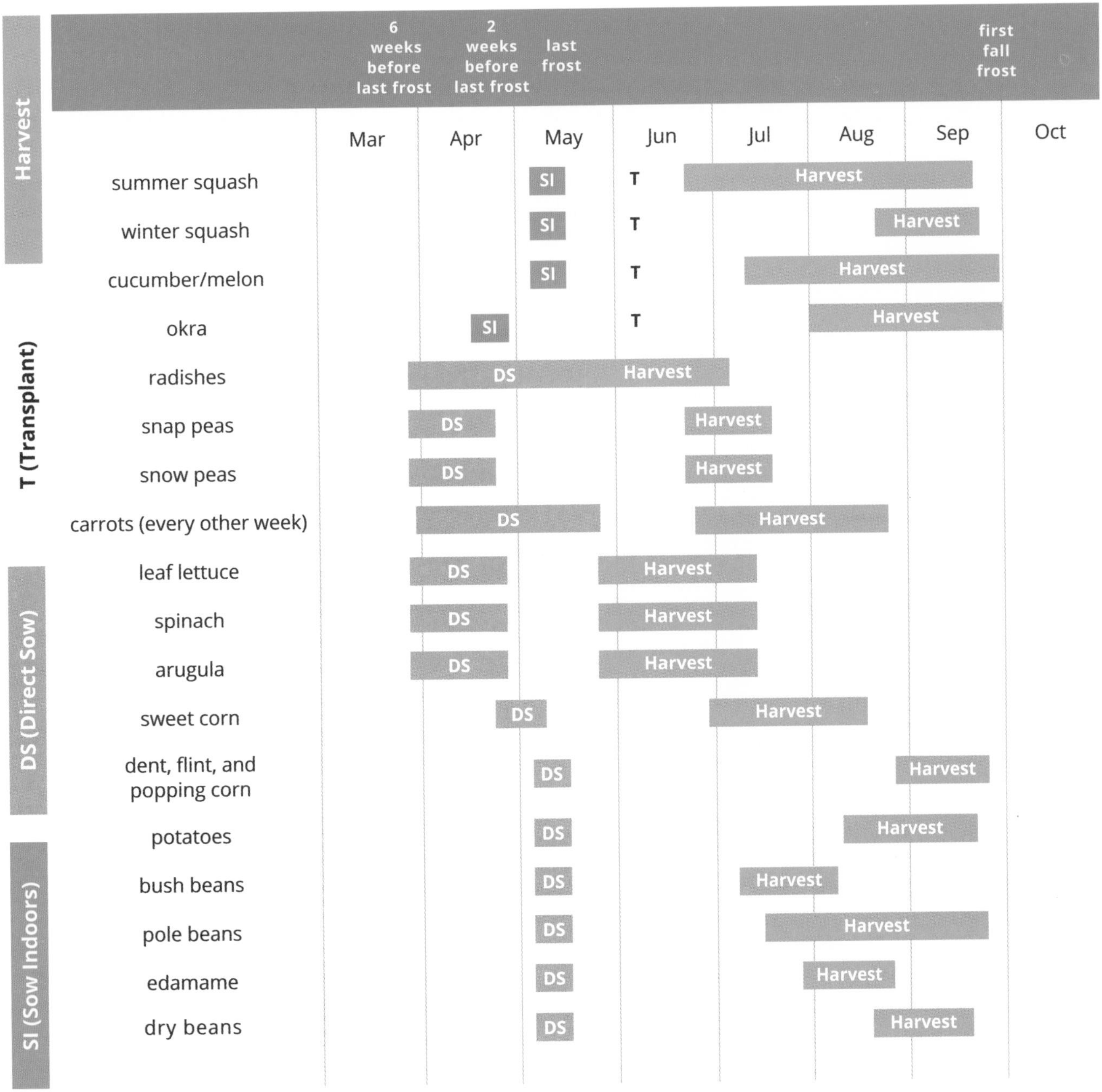

Planting and harvest chart

Left-axis category labels (top to bottom): **Harvest**, **T (Transplant)**, **DS (Direct Sow)**, **SI (Sow Indoors)**

Timeline markers: 6 weeks before last frost (Apr), 2 weeks before last frost (Apr/May), last frost (May), first fall frost (Sep)

Crop	Mar	Apr	May	Jun	Jul	Aug	Sep	Oct
summer squash			SI	T	Harvest	Harvest		
winter squash			SI	T		Harvest	Harvest	
cucumber/melon			SI	T	Harvest	Harvest		
okra		SI		T		Harvest	Harvest	
radishes		DS (Apr–May)		Harvest (May–Jul)				
snap peas		DS			Harvest			
snow peas		DS			Harvest			
carrots (every other week)		DS (Apr–May)			Harvest (Jun–Aug)			
leaf lettuce		DS		Harvest				
spinach		DS		Harvest				
arugula		DS		Harvest				
sweet corn			DS		Harvest			
dent, flint, and popping corn			DS				Harvest	
potatoes			DS			Harvest	Harvest	
bush beans			DS		Harvest			
pole beans			DS		Harvest	Harvest		
edamame			DS			Harvest		
dry beans			DS			Harvest	Harvest	

Key: T = Transplant, DS = Direct Sow, SI = Sow Indoors

ANNUAL AND CUT FLOWER SOWING GUIDE

SI (Sow Indoors) | **DS (Direct Sow)** | **T (Transplant)** | **Bloom time**

- ☾ darkness aids germination
- 🐞 attracts beneficial insects
- 💧 24 hr. soak before planting
- ☀ surface sow / light aids germination
- attracts pollinators and butterflies
- deer resistant

	6 weeks before last frost	4 weeks before last frost	2–3 weeks before last frost	last frost		frost-free growing seson				first frost
	Mar 14	Apr 1	Apr 14–28	May 1–14	May 15-30	Jun	Jul	Aug	Sep	Oct
Mexican sunflower	SI				T			Bloom Time	Bloom Time	
snapdragon	SI¹ ☀	DS	SI² ☀	T¹	T²		Bloom Time	Bloom Time	Bloom Time	
globe amaranth	SI				T			Bloom Time	Bloom Time	
strawflower	SI				T			Bloom Time	Bloom Time	
calendula	SI ☾		DS	T		Bloom Time	Bloom Time	Bloom Time	Bloom Time	
breadseed poppies	DS ☀					Bloom Time	Bloom Time			
sweet alyssum		SI ☀		T		Bloom Time	Bloom Time	Bloom Time	Bloom Time	
nasturtiums			💧 SI		T		Bloom Time	Bloom Time	Bloom Time	
cosmos			SI		T		Bloom Time	Bloom Time	Bloom Time	
marigolds			SI		T		Bloom Time	Bloom Time	Bloom Time	
zinnia			SI		T		Bloom Time	Bloom Time	Bloom Time	
sunflower			SI¹	T¹	DS	SI²	T²	Bloom Time	Bloom Time	

How Early Is Too Early?

It is possible to start plants too soon out of sheer excitement if, like me, you naturally begin to yearn for spring in early February. The process of sowing seeds is as much one of self-care as it is a necessity for pushing your growing season. There is, however, a limit to how well plants fare indoors before heading into the garden. Timing your seedlings so they have strong roots and are hardened off at the right time for transplanting is a bit of an art.

Our earliest attempt at super early brassicas quickly devolved. In early February, we sowed the very first early cabbage and Romanesco broccoli, planning for late March transplanting into the garden. We were shocked by their vigor, and they had to be potted up into 4 ¼-in. newspaper pots within three weeks of germination. Even with the extra room to grow, by the time they were in the ground, the fast-maturing mini cabbage had practically started to head. Subsequently, we've delayed sowing three weeks, and despite weather fluctuations in both directions, this new schedule has yielded harvests of similar size and timing. Mini cabbages and Romanesco broccoli in late May are now a tradition in our family, without the undue stress of sowing them extra early.

Newspaper pots with seven-week-old brassicas ready to be transplanted into their warm low tunnel.

Folding Newspaper Pots

1

2

3

4

5

6

7

8

9

9

10

11

1. Take a full newspaper page and fold it in half lengthwise, or less than half, to make a 12-in. wide by 22-in. long rectangle.

2. Fold in half in the other direction. Be sure to crease well.

3. Fold in half one more time, lengthwise. Crease well and open back up.

4. After opening, fold down two equal triangles that meet in the center crease. Be sure they are close but not overlapping.

5. Take the bottom and fold it in half. The bottom edge now touches the base of the triangles.

6. Fold up again and crease. Flip the pot over.

7. Now, with the back side, fold two rectangles into the same center crease as you did for the triangles on the front side.

8. Repeat steps 5 and 6, and when you double-fold, tuck the paper into the pocket created by the triangles.

9. Fold the pot in half, crease, and open up to form a 90-degree angle. This second to last fold goes along the outside of the pot, creating two triangles on the base of the pot to start shaping it into a square.

10. With the tip facing up and the crease on your right, take the tip of the triangle and fold it to over to the side that is lying flat on a surface. Crease well. This will help you shape the bottom into a square.

11. Gently open the pot from the inside, using that last crease to bend the bottom into a right triangle, and ultimately a square bottom.

EARLY SPRING TASKS

Seed starting	Continue to sow cool season crops; begin to sow warm season crops
	If plants will be indoors more than five weeks, pot up to produce stronger, healthier starts
Warm season crops	Just before last frost, indoor sow cucumber, squash, and melon for transplanting in about three to four weeks
	If direct sowing, sow two to three weeks after last frost
Transplanting	Check soil temperature and long-term forecast before transplanting frost tender plants; take risks each season with extra starts
	Use a tape measure to map out your bed, providing sufficient room to grow
	Add slow-release organic fertilizer to each transplant hole
Direct sowing	Sow radishes, peas, carrots, corn, leafy greens, beets, dry beans, string beans
	Broadcast fertilizer
	Thin seedlings to appropriate spacing to provide minimum square footage necessary
Low tunnels	Vent and close hoops daily to ensure tunnels don't get too warm or cold; either extreme could prove fatal
Perennials	Top dress fruit trees, berries, and other perennial edibles with compost
	Amend soil pH for blueberries as needed based on pH readings
Bed maintenance	Edge in-ground beds to control creeping weeds in high summer
	Top raised beds with a healthy 1–2 in. of compost

A late April harvest from my ceremonial spring equinox sowing means harvest season officially started.

Vernal Equinox Tradition: May the Early Radish Win

Spring has arrived here, at least on paper. We can finally feel a little solar radiation warming our exposed skin. And when it snows (because it snows in spring here), the roads clear to blacktop in a day, rather than the weeks on end of squeaky-cold snow-compacted roads, regardless of how much the sun shines down. Bring on the melt.

The days grow markedly longer, and hope is in the air, although the air is often more than a tinge bitter, occasionally below zero with wind chills. But as the earth tilts and the sun slowly returns north, I know in my heart it's not far, the shift toward the miraculous awakening that is spring. And so I reach for the very first packet of seeds I will direct sow in my garden—on the very first day of spring, which still looks and feels very much like winter for me.

That's when I carry a packet of radish seeds into the garden, tucked into the cavernous pocket of my well-worn winter parka. That hint of spring sees me trudging along in my insulated bibs and boots through what can be several feet of snow, my every move recorded by a stark white landscape. But in the process, my boot steps quickly become worn paths, evidence that I believe. I believe in ceremoniously sowing these jewels of nature that will almost certainly sprout before the end of the month under a warm, protective blanket or two. That day, a wellspring of hope carries my heart. Spring is imminent.

I decided a few years ago that sowing hope, early and often, was the most successful path to gracefully weathering the long drudge of the Minnesota winter. Besides our indoor veggie starts we coddle as early as February, the garden radish has stepped up to provide not only hope, but a tangible, pungent, red orb, long before most any other vegetable.

The early radish is a reliable garden companion, and one I confidently lean on each March for a quick sprig of green in a lingering sea of white. You can almost sense its intrinsic drive in real time as it germinates in the chilliest of the shoulder seasons, swelling to maturity in the blink of an eye. It's always the first to our table by early May. Even without low tunnels, it can be sowed as simply and as soon as you can get your trowel into bare earth, which might be about a month later, in mid-April, if I waited.

The Numbers Game

While arugula and spinach seeds can germinate in soils as chilly as 40°F (4°C), there is a marked hastening of germination as the soil warms. This is generally true for all cool-season vegetables tolerant of chilly soils; while they would eventually break dormancy regardless, it happens much faster as soil temperatures approach and exceed 70°F (21°C). So, while you might sow seeds in low 40-something-degree soil for your mental health as much as anything else, if you make a second pass with more seeds a week or two later, when the soil has warmed to 50°F (10°C), you will see how germination rates, growth, and establishment quickens with more welcoming soil conditions. This second sowing will likely even mature around the same time as the first.

SOIL TEMPERATURES BY GERMINATION RATES

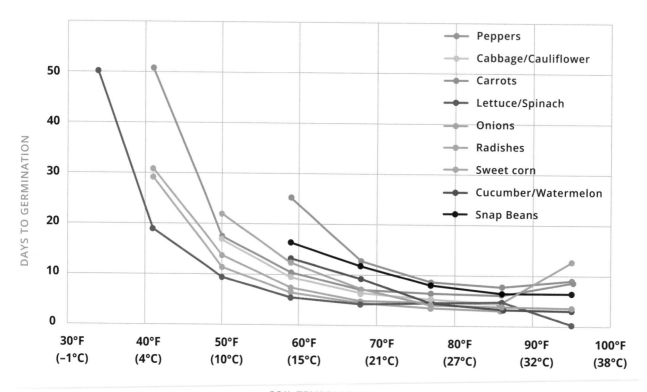

Those late winter tunnels we erect each year hasten soil warming and create a microclimate at least one growing zone warmer. Living a good bit north of zone 5, the national standard for gardening, we are largely beyond the range of so many common perennials like nectarine, sweet cherry, paw paw, cold-hardy Chicago fig, and others. But stretching our annual growing season a good 4 to 6 weeks earlier is well within our reach—and more than worth the effort—for annual vegetables.

MINIMUM VS. IDEAL SOIL GERMINATION TEMPERATURES

Crop	Minimum Soil Tempercature for Germination	Ideal Soil Temperature for Germination
Spinach	32–40°F (0–4°C)	70°F (21°C)
Radish, peas, arugula	40–45°F (4–7°C)	70–80°F (21–27°C)
Onions, leeks	40–45°F (4–7°C)	70–80°F (21–27°C)
Carrots, beets	40–45°F (4–7°C)	70–80°F (21–27°C)
Broccoli, Brussels sprouts, cauliflower, Chinese cabbage, cabbage	40–45°F (4–7°C)	70–80°F (21–27°C)
Corn and tomatoes	40–45°F (4–7°C)	70–80°F (21–27°C)
Beans	50°F (10°C)	70–80°F (21–27°C)

OPPOSITE
This information is interpreted from data compiled by J.F. Harrington of the Department of Vegetable Crops at the University of California, Davis, in 1954.

Early season soil temperatures can vary dramatically. Under a low tunnel, the temperature near the surface may read 45°F (7°C), but if you dig down 6 in., where a transplant's roots will settle in, the temperature might be only 40°F (4°C). I measure my soil temperatures at 6 in. for my brassica transplants, waiting for 50°F (10°C) soils at that depth. Because I harden them off outdoors for a good week while overnight low temperatures hover in the mid- or upper thirties, I know these cooler soils won't shock their roots. For direct seeding quick crops like radish, peas, spinach, and arugula, measuring the top inch of soil at 45°F is sufficient, as their roots will establish as the soil continues to warm.

MAY SOWINGS

Indoor Sowing	Direct Seeding
Vegetables	
Brussels sprouts	Carrots
Celery	Radishes
Kale	Arugula
Heat tolerant head lettuce	Lettuce
Melons*	Dry beans
Cucumbers*	String beans (pole and bush)
Winter squash*	Corn (sweet and dry)
Summer squash*	Potatoes
Fennel	
Flowers (indoor or direct sow)	
Zinnia	
Cosmos	
Marigolds	
Sweet alyssum	
Calendula	
Sunflowers	
Nasturtiums	
Herbs (indoor or direct sow)	
Cilantro	
Basil	
Dill	

Sow in 3-in. newspaper pots to keep roots intact when transplanting.

In addition to all the warm-season vegetables, we ensure our annual flowers bloom early too by sowing frost-sensitive flowers indoors four weeks before our last frost, instead of waiting to direct sow.

Layering Successions

The entire growing season must be mapped out in early spring: your early, mid-season, and late season successions. All planted during or before spring, this layering is what sets you up for extended harvests. Taking the time to think through every layer that's key to your culinary goals now will ensure the ideal harvests throughout the growing season.

Quick successions continue to be planted in early spring and into early summer. These fast-maturing vegetables fill baskets quickly, but most will fade away come heat. Cilantro, lettuce, radishes, arugula, spinach, and kohlrabi are excellent quick successions to continue to sow, either indoors for a head start, or directly into the garden a few weeks before your last spring frost. Their accelerated growth is why these plants make such excellent succession plantings for continuous harvests. These low growing vegetables are always interplanted in our garden.

Midseason successions are something of a mixed bag during spring. Some we pot up to prepare for warmer days, including tomatoes, peppers, and eggplants. Others are remarkably resilient, and can be direct sown in slightly cool spring soil. These include sweet corn and snap beans. Spring is also a fantastic time to sow your generalist successions: carrots, beets, heat-tolerant brassicas, and herbs. Indoor sowing cucurbits (cucumbers, summer squash, and winter squash) in 3-in. newspaper pots around your last frost will give these plants a head start on summer; direct sowing would be delayed two to three weeks until soil temperatures top 70°F (21°C), which is early June for me.

At this point, late succession vegetables are also already on their way, and in spring, we add these final layers. Potatoes are planted right around last frost and sit tight, anchoring their bed all summer long. Leeks have been holding tight in their pots, and are also headed out to their permanent summer and autumn home in the garden. Brussels sprouts, dent corn, popping corn, and dry beans are among the last late succession vegetables we plant, the former indoor sowed and transplanted, and the latter three direct sowed.

Making the effort to plant your garden during the early spring shoulder season maximizes your growing space. It not only provides hearty, nutritious food ahead of the usual summer vegetables, it also turns a bed over early enough in the growing season to get in one or more successions after it. Traditionally, this bed might not be planted until late spring, but by harvesting the first crops starting in May or early June, you can easily follow that first succession with any number of fantastic mid-succession crops like beans, carrots, beets, sweet corn, summer squash, or cucumbers.

Timing shoulder season plantings in this way will take your productivity to new levels. These first harvests are the essence of succession gardening in action, maximizing your growing potential, producing as diverse a cornucopia of food as possible, beginning as early as possible, all the way through to the dark and short days of the dormant season. You will quickly appreciate how much more you can grow as a succession gardener in spring, wherever you garden.

Annual flowers are the best of both worlds: added nectar and maximum flexibility year to year as you redesign and plant your garden each spring, trying new combinations, and learning as you grow. Benary's Giant zinnia are an annual must-grow for me.

ANNUAL FLOWER SUCCESSIONS IN THE FOOD GARDEN

In addition to perennial gardens and clover-infused living lawns, we add another layer of flower succession to our food garden, for all the same reasons. The beauty of flowers parades across the seasons, providing an extra buffet for pollinators throughout the growing season, in very close proximity to our most prized harvests from the food garden. You can grow a really good food garden without flowers, but a food garden infused with flowers becomes something exceptional—a downright edible paradise.

Some of the flowers discussed in this chapter, though beloved annual companions in our chilly zone 4 garden, may self-seed more readily in warmer climates. While few self-seed here, I suspect the frigid plunge of winter reduces the seed bed in our food garden. And I am more than OK with that. I prefer to build a garden literally from the ground up each spring. Annual flowers are the perfect accompaniment to our annual vegetable garden because they invite creativity and personal growth as we play with the infinite interplanting possibilities between food and flowers.

As with the food we grow, we promote diversity further by growing many varieties of flowers. By including flowers in our vegetable garden, we attract a robust population of insects to help curtail pest infestations and aid in pollinating the flowering vegetables. Some annual flowers are favored by pollinators, while others attract insects that are beneficial in other ways. Planting as many different types of flowers as can fit in your garden effortlessly supports this cause—plus, adding flowers to your vegetables will bring you joy. Let's discuss the primary annual flowers we grow in and around our food garden.

Flowers within flowers, it's no wonder the sunflower is a magnet for bees.

Meet the Composites

In my garden, composites are to flowers what the brassicas are to vegetables: indispensable. This diverse and enormous plant family (over 20,000 species worldwide) includes my most beloved annual companions, and my garden would never be complete without them. Cosmos, zinnia, marigold, sunflower, and calendula are all members of the composite family. Sometimes called the daisy or sunflower family, this brood also includes edibles such as lettuce, dandelion, and artichoke, as well as many native prairie plants like coneflower, aster, and goldenrod.

As the word "composite" suggests, the flowers of this plant family are an amalgamation of more than one flower. Each bloom we enjoy at the macro level is comprised of dozens of small, often inconspicuous, counterparts in the center known as disc flowers, which open from the outside in. Surrounding the disc are ray flowers. The petal of a ray flower is actually a modified leaf, and each ray correlates to a single "petal" around its flower. Each composite flower actually encompasses hundreds or even thousands of flowers.

It's no wonder this plant family is both prolific and beneficial, and these plants are some of the most nutritious pitstops we can add to our summer gardens. Most on my list hail from Mexico and Central America, and they're the mainstay of my main and late summer annual flower successions. They are useful both as foundation plantings and interplanted among the vegetables, bringing diverse height, bloom time, and color to our food plots—and sometimes onto our plates.

Calendula

Cheerful and frequented by pollinators, calendula, also commonly called pot marigold, is compact and aromatic. Calendula flowers profusely early in the season, fading to the background when summer heat sets in, and other flowers have arrived to fill the garden. It then roars back to glory in late summer and early autumn, happiest to make an encore after a good haircut in August, flowering late into the shoulder season. It makes for a reliable garden friend, particularly in cold climates because it can take a light frost. My early experience with calendula, however, is a cautionary tale.

Calendula blossoms embody understated elegance. But don't be fooled: these are prolific self-seeders in the home garden, so plan accordingly.

Many varieties of calendula offer similar stature in the garden, but I am partial to the simple floral structure of Resina and its fragrance.

I was first drawn to the simple, daisy-like flowers of the variety Resina, a trusted medicinal variety. My first year growing it, I enthusiastically harvested, air dried, and infused my dried flowers in oil. I made healing salves, given as gifts grown from seed. But later that very first year, I noticed dozens and dozens of calendula seedlings sprouting in the cracks of my driveway. I quickly understood how easily this plant could get away in a single season, establishing a hefty, potent seed bed. After that first season, I planted fewer, and was proactive and strategic in deadheading to limit self-seeding.

I became so efficient at deadheading and removing plants, but without saving seed, that Resina eventually disappeared from the garden. While I had a few other varieties in my deep seed stash, I lamented my shortcomings, frustrated by the lack of locally available seed to purchase. Then, in late summer, self-sowed calendula sprouted around the gardens. On a hunch, I let them flower. Lo and behold, those previous years' dropped blossoms produced my beloved orange Resina calendula, restoring my seed supply and delighting my senses. Sometimes a messy garden is just the teacher we need.

I tuck in calendula here and there around the garden. As a low-growing medicinal herb, it has a strong aroma that lifts my mood, and I often seek it out for a grounding dose of joy. I love to pair it with compact plants, especially cabbages and eggplants, so it can pop cheerfully between the broad leaves. I also like to plant it at the edge of a vertical structure, next to beans or cucumbers, where it can thrive in the partial shade of the vines. What I love most about it is how long it holds in the garden in late fall, after my more tender flowers have succumbed to our first freezes.

Zinnia

Zinnia are a must-grow for me. Sometimes we grow something to honor our ancestors, and zinnia are my ancestral plants. They are the flower that links me back across the generations to my great grandmother, the matriarchal gardener. They adorned our kitchen countertop every summer growing up.

Zinnia anchor our garden. Here, I interplanted meadow blazing star with orange Benary's Giant zinnia, as both flowers attract monarch butterflies.

Benary's Giant is my favorite zinnia variety. Three months of constant blooms in an array of hues, they easily reach 4 ft. in height, and anchor our food garden with their utilitarian pops of color. They make excellent cut flowers, doubling as dependable beacons for the insects from the height of summer until the first hard freeze. When sowed indoors four weeks before last frost and transplanted just after it, zinnia will flower from late June into fall, depending on when the first killing freeze arrives. If there is room for only one sprawling flower in your food garden, prioritize space for this multi-tasking powerhouse.

ZINNIA SPECIES AND COMMON VARIETIES

Zinnia elegans

- Benary's Giant
- Northern Lights
- Oklahoma
- Polar Bear
- Queen Lime
- State Fair
- Senora
- Thumbelina
- Whirligig Mix

Zinnia haageana

- Jazzy Mix
- Persian Carpet

Not all zinnia possess the same pollinator magnetism, though most are prized watering holes for pollinators and butterflies alike. Searching for a compact zinnia many years ago, I fell in love with the Profusion series, a past winner of the All-America Selections (AAS) trials. These zinnia fit the bill, remaining tidy, and flowering profusely without the need for pinching to stay beautiful all season long. I adored them for a few seasons before I noticed it: this double-blooming hybrid is largely passed over by pollinators in favor of other flowers in the garden.

A wave of guilt washed over me once I understood that my desire for a tidy flowering border had won out over the ecological benefit of this garden space. Do I still grow this flower? Admittedly, I do, although I've reduced the number of plants. I do so knowing full well that it pales in comparison to other zinnia in my garden. Nevertheless, a small planting of flowers for my visual pleasure won't detract from the vast offering of nectar and pollen elsewhere. When you grow along with your garden, you will tune in to these observations, and can decide for yourself what works best. For me, that balance tips in favor of feeding our insect populations, and my preferences in flowers have admittedly grown to accommodate their preferences.

Since it's one of the taller flowers we add to the food garden, I plant zinnia almost exclusively in dedicated beds instead of interplanting with vegetables. Typically planted on the ends of rows, they quickly consume the space they invariably need. Thriving at the edges of beds, they innocuously stretch out into the pathways instead of into the neighboring food, where they would compete for the precious sunlight the vegetables need to successfully produce. Zinnia fill out quickly, branching and becoming lush, bushy plants. The more you prune, the more they will flower. They are a source of daily joy and peace, and rival our most beloved foods we also grow from seed.

Sometimes we choose to grow a flower that pleases us more than the pollinators. Profusion zinnia is that flower for me.

Cosmos

Nothing says old world charm more than the paper thin, open arms of a cosmos blossom. Their saucer-like flowers are as delicate as their wispy foliage, and add a touch of whimsy and softness to the food garden. Cosmos come in a wide range of colors and sizes, and make excellent cut flowers. There are two species of interest: *Cosmos bipinnatus* boasts pastel tones and fernlike foliage, and *Cosmos sulphureus* more fiery blooms with foliage closer to that of a marigold. They bloom similarly to zinnia in my experience, if somewhat less profusely, although that may be due to my lack of deadheading. Once they get going, they are a beautiful summer garden companion.

COSMOS SPECIES AND COMMON VARIETIES

Cosmos bipinnatus (white, pink, fuchsia)
- Cupcake
- Double Click
- Rubinato
- Seashells
- Sensation
- Versailles

Cosmos sulphureus (orange, orange-red, and orange-yellow)
- Bright Lights
- Buddha's Hand
- Sulphur

Cosmos is another stellar food garden addition, attracting many types of bees, including this honeybee, who gratefully feast on them.

I add cosmos sparingly to the food garden, partly because it grows so tall, but more importantly, it seems prone to wind stress. More often than not, the corn and kale stand tall, but the cosmos is the single plant that blows over during a gusty storm. Because I focus my trellising and support efforts on our food production, my flowers must be self-sufficient. Interplanting cosmos among zinnia and strawflower provides adequate protection from high winds. Pinching the plant back early on also helps it grow bushier, thus reducing potential wind damage. However, knowing these tricks and finding the time to implement them with regularity are two entirely different things.

Tangerine Gem marigolds are a delight with their small, simple flowers that appear profusely in the height of summer.

Marigolds

Cheery and fragrant marigolds are another childhood tether that makes a substantive food garden companion. At a young age, I was taught how to pinch out spent marigold blossoms, encouraged to do so by my busy mother. Like their cousin, calendula, marigolds are highly fragrant. After many summers of pinching, I developed mixed emotions, the aroma flooding my mind with memories of being put to work when I'd rather have been playing. Because of this, marigolds didn't make their way back into our gardens until more recently. Like cosmos and zinnia, there are several types of marigold.

Marigolds are generally compact, and they flower profusely once mature. Purported to deter pests and invite pollinators, they are also some of the more delicious edible flowers in our garden. I'm partial to some species over others, and that's merely personal preference. I prefer the look of a simple flower, so I grow signet marigolds more than other types, as their blooms are reminiscent of a miniature sulphur cosmos.

MARIGOLD SPECIES AND COMMON VARIETIES

Signet marigolds (*Tagetes tenuifolia*)
- Lemon Gem
- Red Gem
- Tangerine Gem

African or Aztec marigolds (*Tagetes erecta*)
- Crackerjack
- Giant orange
- Giant yellow

French marigolds (*Tagetes patula*)
- Durango
- Lemon Drop
- Petite
- Sophia

Marigolds work well as border plants, or under any vining crop, but I plant marigolds between my peppers annually. Between our pepper plants, they form a beautiful mass, providing shade for the ripening peppers and creating a green mulch, keeping the soil moist and cool and the weeds at bay. That said, as a low-growing flowering plant, marigolds are a willing companion nearly anywhere in your food garden.

Blooms of Mexican sunflower are the orange to beat in the summer garden.

Mexican sunflower needs a large, preferably open space so it can spread out, as evidenced by this massive plant start. When given proper square footage, it can exceed an area of 5 cu. ft., a veritable annual shrub.

Mexican Sunflower

A close cousin to zinnias, Mexican sunflower (*Tithonia rotundifolia*) is a large, shrub-like annual that easily tops 5 ft. before the season closes. Attracting beneficial insects, butterflies, and even hummingbirds, it's a must-grow late season flower in any annual pollinator or auxiliary garden. Also called torch sunflower, this is a plant that will thrive in your auxiliary garden—it is not easily integrated between vegetables, for its towering stature is sure to throw shade on your prized food, muscling its way to dominance.

We give this summer superstar extra time indoors, starting several weeks before zinnia, at least six weeks before our last frost. Direct seeding is always an option, and it will simply delay flowering. In cold climates like ours, every single growing day counts, so we tend to start our flowers indoors for maximum color when summer's heat arrives. To enjoy this pollinator magnet for as long as possible, I provide extra space for them indoors, potting them up into larger pots before they are garden-ready. They're worth the extra care for the food they provide the pollinators.

Soraya branching sunflowers, a true feast for everyone, delight us for a good two months, feed the bees, and then become forage for goldfinches.

I work around sunflowers' sensitive systems by sowing seeds in generous 3¼-ivn. newspaper pots, which get planted directly into the garden around the same time I sow my melons.

Sunflower

These sun worshippers are a late season edible flower succession. Worth the wait, sunflowers bring plenty of food and a sunny disposition to pep up the tired garden and gardeners alike. Sunflowers come in both single-stemmed and branching varieties. While single stems offer a single burst of color on that long, sturdy stem, that's all you get—but with varieties like Russian Mammoth that flower for several weeks straight, it's like a succession of flowers within flowers. Branching varieties offer a steady stream of floral succession, their own staggered planting, much like the branching and bushy habitat of zinnia and calendula, though markedly more abbreviated.

For the most potent pollinator benefit, sunflowers should contain both nectar and pollen. Breeders have made great strides in offering lovely colors and petal textures, but with that, the nutritional value for insects has diminished to

Late summer in the garden and the sunflowers, cosmos, and zinnia are commanding the gardeners' attention.

varying degrees, not unlike what has happened with cultivars of native perennials. When purchasing, avoid seed with the term "pollenless"; they will be useful to many beneficials and butterflies who can feast on the nectar, but don't produce a complete meal for our essential garden workers, the bees.

I take a hybrid approach to my sunflower successions. Unlike the rest of my annual flowers, who are only sown once a season indoors, generally four weeks before our last spring frost, I sow sunflowers indoors and direct seed them into the garden. Sunflower seed packets recommend direct seeding because their roots are sensitive to disturbance. I combat that by growing them in compostable newspaper pots that can be planted right into the garden, or carefully removed and composted at planting. While I find my indoor sowed seedlings more vigorous, I can't help but toss a few seeds into open spaces as insurance for late summer and early autumn cheeriness.

I indoor sow my branching and single stem sunflowers just before our last spring freeze. I sow single stem varieties once, but sow the branching varieties again in three to four weeks, direct sowing them as late as early July. It is the only annual flower I actively succession plant in our garden, in part because its floral life cycle is more abbreviated, but also because it's a stunning flower in a vegetable garden. Sunflowers will self-sow too, adding random spikes of cheeriness where you choose to let them take hold. I love knowing that they will reliably arrive in late summer, and strategically position these large foundation flowers on the western side of the garden so their sunny disposition will cascade across the entire space.

SUCCESSION OF ANNUAL FLOWERS

First Frost

	May	June	July	August	September	October
Sunflower			🌻	🌻🌻	🌻🌻	
Zinnia			🌸🌸	🌸🌸	🌸🌸	
Cosmos			🌼🌼	🌼🌼🌼	🌼🌼	
Marigold		🌺	🌺🌺	🌺🌺	🌺🌺	
Snapdragon	🌷	🌷🌷	🌷🌷	🌷🌷	🌷🌷	🌷
Calendula	🌼	🌼🌼	🌼🌼	🌼🌼	🌼🌼	🌼
Sweet alyssum	🌸🌸	🌸🌸	🌸🌸	🌸🌸	🌸🌸	🌸🌸

Sweet alyssum flowers all season long, past first frost. Calendula and snapdragon bloom less in the heat of summer, but flower past first frost. Marigold, cosmos, and zinnia all bloom from late June until first frost. Sunflowers are later to start, and also don't last past first frost.

Companion Planting Powerhouses

Out beyond the vast family of composite flowers are two very special garden companions. Sweet alyssum and nasturtium are a pair of indispensable flower companions in the food garden. My annual vegetable garden would not be complete without them. Both are edible, attract pollinators and beneficial insects, and create a living mulch in the understory of the garden beds.

Sweet Alyssum

Sweet alyssum is a dynamite annual flower that is quick to mature and provide nectar, yet lasts the season too. A relative of our beloved culinary brassicas, it is my go-to flower for our food garden because of its immediate benefit to our ecosystem. Flowering in less than two months' time, when started indoors ahead of transplant, these plants will flower very early in spring, kicking off the season for beneficial insects, setting the table for a long and delicious feast that will last for the next five months.

Sweet alyssum is a nectar source for beneficial insects and pollinators from early spring through late fall. It generously blossoms in our vegetable garden by mid-May, grown from seed in early April. Six weeks to its first blooms, the tiny

Before the sweet alyssum even reached the garden, this hungry, early season Milwaukee miner bee located and feasted on the plants hardening off on the deck.

seeds are packed with a penchant for thriving. Their delicate foliage, massing growth habit, cold tolerance, and low profile makes sweet alyssum my number one flower for bordering any and all vegetable beds.

This plant attracts hoverflies, a vast group of predatory insects whose larvae feed on aphids and other garden pests. These are adorably diminutive beneficial insects you want in your garden. The more, the merrier, and in our garden, their numbers keep increasing, which means at least two things: first, we are providing enough nectar to support the adults; and second, we are providing enough food for them in the larval stage to sustain their entire life cycle.

Sweet alyssum's other major benefit is that it's a green mulch. Because it spreads more horizontally than vertically, it makes a perfect addition to nearly any vegetable planting, assuming you left a little sunny space at the edge of

Once established and happy, nasturtiums will spread out and ramble as they are doing in our espalier orchard. Interplant this one with your vegetable space needs in mind.

the bed. As it grows and spreads, providing pollen and nectar to beneficial insects, it does double duty as a weed suppressor, smothering and shading out weed seedlings that may be trying to germinate, all the while holding in moisture. It really is that good—interplanting made easy. I reach for this flower most often in the vegetable garden.

Nasturtium

In a botanical class all its own, nasturtium is a rambling creeper revered by the most enthusiastic gardeners. Its foliage and flowers are unique and edible, and this warm season plant prefers your less fertile soil, where it will flower profusely. Nasturtiums are added as much for a splash of color and texture as for pollinators. It is also purported to be a useful trap crop and pest repellent, so its benefits likely extend beyond sips for pollinators and tasty, pungent flowers for humans. And if that weren't enough, hummingbirds frequent its flowers in late summer.

I plant most of my nasturtiums, like my zinnia, in spaces dedicated to flowers within my garden. These are not places we're producing as much food, nor where interplanting would impede food production, such as in the understory of our fruit trees. This is because nasturtiums are energetic ramblers. I've wrangled interplanted nasturtiums among tomatillos, and the nasturtiums always win, vastly reducing food production. So I allocate specific areas for these peppery flowers in the margins, where food is not as prominent. They make an excellent auxiliary garden plant too.

NASTURTIUM VARIETIES

- Alaska
- Bloody Mary
- Empress of India
- Fiesta
- Jewel
- Ladybird Rose
- Orchid Cream
- Tall Trailing

Planted in the auxiliary annual flower garden along our driveway, adjacent to one of our planted prairies, Bicolor Rose globe amaranth makes a strong border plant, defining this pollinator garden with its lush foliage and long-lasting flowers.

Snapdragon flowers surreptitiously greet us among a bed of vegetables, where they add old world whimsy.

Honorable Mentions

Once you have added a few pollinator magnets to your garden, any other flower is a bonus. My list of favorite flowers includes those of our establishing perennial prairie, which provides the right plants at the right time. Additionally, the food garden is graced by more than just the flowers already discussed. Here are a handful of other wonderful flowers to consider adding to your landscape.

Globe amaranth (*Gomphrena* cvs.) is a beautiful annual flower. Its compact form and everlasting flowers make it a wonderful companion in our garden. I love to use it as a border plant in front of a mix of zinnia, snapdragon, poppies, and cosmos. It tops out under 24 in., and creates a thick mass by the middle of summer. Globe amaranth takes a little longer to establish, so this is one I always sow indoors a good bit ahead of the rest. Once in bloom, it requires zero effort or deadheading, a truly low maintenance annual flower.

Snapdragon (*Antirrhinum* cvs.) is another old timey flower I have recently fallen in love with as a vegetable interluder. Its spiky form brings a specific visual interest to the garden. The varieties I've grown to date are also quite compact; they complement just about any vegetable without competing for light or shading anything out. They are also tolerant of the shoulder season, and I love them for those early and late season color pops. Snapdragon varieties self-seed readily, so that's definitely something to take into consideration. This flower is happy direct seeded in early spring.

Hungarian Blue breadseed poppies and our American elderberry shrub blooming together, both fantastic pollinator magnets in the auxiliary garden.

Honeybees methodically circle the unfurling blossoms of strawflower, and I adore observing their fastidious work.

Breadseed poppy (*Papaver somniferum* cvs.) is a delicate and striking addition to the food garden. As its name suggests, it is the source of poppyseeds for baking. Breadseed poppies are most often direct seeded because their roots are sensitive to disturbance. They also do best with stratification—that is, a period of cold before germination—and require light to germinate. All these factors are most effortlessly achieved when you scatter seeds into a prepared area in late winter or early spring. They do require more space than other flowers, which is a consideration when placing them. Breadseed poppies flourish in my auxiliary flower gardens outside the main food garden.

Strawflower (*Xerochrysum bracteatum*) is another everlasting that graces our summer gardens. I happened upon it in a seed catalog and fell in love with the simplicity of its flower. A native of Australia common in cutting gardens, the straw-like texture of strawflower's bracts (colorful, modified leaves) adds a unique element to bouquets. The flower is a magnet for pollinators, and a joy to observe at every stage of inflorescence. These plants grow a good 3–4 ft. tall, so I prefer to plant them in a dedicated annual flower bed, either in the food garden or in our auxiliary gardens—sometimes in both. I sow indoors a bit earlier too, around the same time as globe amaranth, because they are slower to germinate and establish.

RECOMMENDED MINIMUM FLOWER SPACING

Flower	Spacing	Location	Tips
Calendula	6–12 in.	Interplant with more compact crops including brassicas, lettuce, eggplants; happy to grow along edge of bean trellises and as border plant	Deadhead often during growing season to help manage future seed bed; interludes happily with everyone
Zinnia	12 in.	Ends of beds, plus dedicated pollinator beds, for large swaths of color	Branching form; more room is better
Cosmos	12–18 in.	Interplant with vining crops, bed edges, auxiliary pollinator gardens	Benefit from support from neighboring vines or flowers
Marigold	12 in.	Interplant with peppers, beans; makes excellent bedding border plant	Masses into mounds by midsummer, so consider giving extra space
Mexican sunflower	24 in.	Auxiliary pollinator garden superstar	Give plenty of space
Sunflower	12–24 in.	Delineate plantings with sunflowers on west side; plant in rows for living hedge	Branching types bloom longer but need more space
Sweet alyssum	6 in.	Interplant under tomatoes, in between brassicas; use as bedding border plant	Will quickly form carpet at 6 in. spacing; cascades nicely in raised bed
Nasturtium	8–12 in.	Lush border plant under tomatoes, cucumbers, or squash; plant under fruit trees as edible understory and beneficial insectary	Trailing and climbing types need support; grow mounded types for rounded tufts of foliage with pops of color; will often cascade out of raised beds

Herbs with Benefits

Many herbs make fantastic garden additions, their flowers sought after by beneficial insects and pollinators. Most of these are beloved culinary herbs, and some we celebrate for their flowers, though their foliage is prized most often. Either way, letting a few go to flower, especially in a small, urban setting, is a simple yet effective offering to the insects.

Basil comes in so many shapes and sizes, with over 100 cultivars worldwide. Eager to flower once summer heat lingers, it's a plant for all to enjoy. My favorite basil for aroma is tulsi, also known as holy basil. It makes a great tea and is renowned as a medicinal plant dating back centuries. The intrinsic pull of its aroma confirms, for me, there's something special about this one.

For culinary purposes, we lean on more conventional basil varieties. We love Genovese basil for traditional pesto. Once we stock up our pesto rations, we abandon pinching, allowing it flower for the duration of the season. Opal

Tulsi basil does it all: its highly aromatic foliage soothes your senses, its leaves can be dried for tea, and it's yet another layer of succession for the pollinators.

(or purple) basil offers visual variety in the garden, though truthfully, for us, it behaves more like a flower than herb, and we interplant it with our tomatoes and sweet alyssum. Thai basil is equally easy from seed, and widely used in southeast Asian cuisine. It adds a hint of anise to your dishes, and is great to have on hand for spring rolls and curries.

Chives are an early summer pollinator magnet. If that weren't enough, they are deer resistant, and make a fantastic garden border to repel those persistent grazers. Incredibly easy to establish, chives will flourish just about anywhere. We inherited a chive patch when we arrived here, and behind my turned back, it has spread quickly over the past several growing seasons. I now take a more active approach to deadheading the flowers after the bees have had their fill.

Last but not least, dill is my favorite herb in the garden. In spite of our deep admiration, dill is an herb that continues to evade abundance here, and as such invokes both curiosity and respect. The prized flower heads are so rarely in

Dill attracts hoverflies with a deliciously accessible flower structure similar to sweet alyssum.

endless (or even sufficient) supply exactly when I need them for pickling. We collect seeds to use in lieu of dill flowers, though it's just not the same. Besides being a kitchen staple, dill is host to swallowtail caterpillars, and its flowers make welcome landing pads and rendezvous spots for predatory insects like hoverflies.

We sow dill indoors and welcome self-seeded plants when they grace the garden. We manage to fill our quarts of pickles one way or another with either flower heads or saved seed. Let's acknowledge that this plant is not known as dillweed by coincidence: if you let those seed heads drop, you will be blessed for years with dill sprouting up all over the garden. If that's your style, embrace it.

Flowers, flowers, everywhere. If you are growing vegetables with flowering parts that you eat—and you probably are—then your garden will directly benefit from finding space for more flowers. Adding flowers in with your tomatoes, squash, beans, corn, peppers, and eggplants will benefit everyone. Situated near your food garden, your auxiliary gardens will boost insect diversity and increase pollination of your vegetables. The more diversity you bring into your garden across the seasons, the more robust the insect population—and the more balanced your ecosystem. So, what will it be? Annual flowers interplanted in your vegetable garden? Flowers mixed into your lawn? An annual auxiliary garden? Hopefully, it's all of the above.

Interplanting every nook and cranny of the garden, including vertical structures, results in breathtaking garden views—and a highly productive food garden.

THE ART OF INTERPLANTING

Interplanting is nature's most implemented strategy. Any amount of time spent communing with nature will elucidate this fact. Plants entangle in a veritable battle for leaf area aboveground while negotiating water and nutrition uptake below. Some plants tolerate low light levels, while others reach for great heights. How you apply this natural tendency to your food garden has a significant impact on what you reap come harvest.

Interplanting is crucial in short growing season climates and urban gardens alike. It is the means to achieving multiple successions in a single bed, maturing concurrently. It is the path to producing as much as possible, and making the most of every square foot of available soil. Even with a rambling garden, interplanting bestows beauty, cultivates joy, attracts pollinators, and increases productivity.

Interplanting Best Practices

Just how far can you squeeze plants in and still yield well? It all depends on who you are pairing up, what their growth habits are, how their leaf areas complement or compete with one another. Part of the process is how comfortable you

A raucous garden party in the height of summer is made sweeter with strategically interplanted flowers like these Resina calendula.

are with embracing chaos—the potential to sift through the subtropical food jungle you've cultivated, for example, to harvest the prized fruits of your labor.

When you intensively interplant, you eventually reach capacity both above and below ground. Taller, faster growing plants with larger and denser leaves quickly dominate the space, creating a canopy. Ideally, lower growing interplanted crops mature before this canopy closes, otherwise they will often compensate by stretching their stems or dropping their lower leaves (this is known as self-pruning). The ideal interplanting maximizes health and vigor, and brings diversity to soil health without causing plant stress. This can be achieved in several ways.

Pair evenly growing plants together

Interplanting vegetables with similar growth habits and mature heights is perhaps the most straightforward strategy. Blend together brassicas such as cabbage, broccoli, cauliflower, and Romanesco, all of which top out around 2 ft. Interplant these with head lettuce, spaced appropriately, and you have a steady stream of food from a single space for several weeks or more. As the planting matures, outer leaf margins eventually reach one another, closing the canopy.

Mind the gap

Broccoli and head lettuce close in on one another as the lettuce reaches maturity. Once the lettuce is harvested, the broccoli will quickly fill in, closing the canopy a second time.

The most reliable fast-growing crop is radish—and for good reason. Radishes fill the baskets and occupy the space that are not yet needed by their slower growing neighbors, at moments (in spring especially) when plant growth is sluggish at best. An indispensable interplanting partner, use it often in the shoulder seasons. I hardly ever give radishes their own dedicated garden space, save the larger daikon and watermelon radish, whose stature commands it. Quick radishes are always planted between spring crops, including peas, cabbages, broccoli, kale, and cauliflower.

The Power of Sunlight

Be vigilant in selecting the sunniest spot in your yard for growing food. For my family, this meant removing a few trees before building our garden. With the prospect of a south-facing aspect, it was worth the effort to take down a few trees instead of trying to grow under the canopy of black walnuts in another area of our property. We replaced the sacrificial trees with thirteen fruit trees in the new garden space.

In a small space, light is key. Plan your space effectively so you don't shade your beds out. Tallest plants go in the back, ideally on the north side. This way their shade falls outside the garden, to their north, ideally in the paths. Shorter growing plants can happily thrive in front of the taller plants, or in beds on the south side of the garden. Make every square inch count.

The smaller the space, the more creative you will want to be with your edges. What kinds of plants can occupy your edge? Can you allow something to trail off out of the garden? A squash would happily roam into a pathway. If space is limited, how truly necessary is walking unencumbered on that path in late summer versus harvesting your own squash for fall and winter meals?

Be creative. Where can you borrow space, and how can you strategically stretch your space beyond its physical capacity to add plantings and increase your successions?

Be clear about what will become dominant plantings

Interplanting invites two plants to compete for dominance. One will win. It's just their nature. It can be a joy to watch, yet tinged with regret, and seasonal experiments will earn you immeasurable knowledge. By the end of the lively growing season, the plant with the most leaf area wins. Two poorly suited vegetables may end up competing for leaf area; one will eventually be shaded out. Not all pairings work well, and the joy is exploring which ones work well for your garden.

A red cabbage is swallowed by the overstory of dominant Brussels sprouts closing in around it before it had a chance to form a robust head. This is a poor interplanting pairing.

Strategic early season interplanting reaps benefits all season long. The area where an indeterminate main season succession crop will go (tomatoes, for example) is an area that could be utilized in the short term using low-growing, fast-maturing plants. Carrots can be sowed before tomatoes go in, maturing just as the tomatoes begin to dominate the canopy. Lettuce, spinach, or radish would also work well. This is a wide open space for you to explore and wander, mix and match to your climate, taste buds, and growing style.

FAVORITE FLOWER AND VEGETABLE COMBINATIONS

Main Season Overstory	Potential Interplanted Understory
Beans	Lettuce, nasturtium, calendula, sweet alyssum
Tomatoes	Herbs, lettuce, carrots, sweet alyssum
Peppers	Herbs, lettuce, marigolds
Brussels sprouts, cabbage, broccoli, cauliflower	Lettuce, radishes, beets maturing before canopy closes
Popping and dent corn	Beans, squash, sunflower, zinnia, cosmos
Summer squash	Herbs, lettuce, radishes, arugula, early beets, calendula, nasturtium, snapdragon
Cucumbers	Herbs, lettuce, cosmos, calendula, sweet alyssum, snapdragon
Potatoes and onions	Not recommended; give these crops their own space without interplanting

Choose plants with complementary foliage

Interplanting is about maximizing your successions across the seasons; selecting plants with foliage that complement rather than compete is a prerequisite for success. The primary foundation plant with denser, more dominant foliage will take center stage. These are often your main vining crops in the height of summer, such as tomatoes, cucumbers, squash, and beans. Select a secondary plant with softer foliage, whose presence will enhance the planting without detracting from the productivity of the primary producer.

Do the math

The wispy and whimsical foliage of cosmos pairs perfectly, going vertical without taking up a lot of leaf area or shading out its cucumber neighbors.

Read up on each variety and pairing you consider. Take into account its days to maturity, height at maturity, and how long it will be productive. Indeterminate plants (those that produce indefinitely once mature) and vining crops will last as long as warm weather persists. You may be able to make unlikely pairings based on how you time your season. Capitalizing on the understory of a waning tomato bed with frost hardy vegetables, for example, will add a succession to that bed, keeping it productive long after first frost.

Consider plant structure and form

Interplanting means occupying space both horizontally and vertically. When that's done well, plants thrive. Pairing two rambling plants is a missed opportunity for vertical gardening. Likewise, vertical gardening without an understory planting underutilizes space. Be it vegetables or beneficial flowers, mix and match plants so their structures complement one another.

Just do it

The permutations are truly endless with interplanting. Possibilities await that are unique to your planting style and aesthetic. Embracing your culinary curiosity and the idea of seasonal eating will generate ideas for interplanting too. We go without green salads for several weeks each summer, for example, and instead embrace the hot season crops, our diet reflective of what's in season. Interplanting will give you the chance to really focus on what your seasonal menu can provide.

Another failed interplanting, the rambling (trailing) nasturtiums decided the tomatillo was its trellis, quickly engulfing the planting, overwhelming the area and reducing productivity, although it was a colorful and fragrant sight to behold.

Growing Up

Vertical gardening is the single best way to grow more. Vertical elements can be temporary or permanent structures. Foundation plants can become structure for interplanted vining plants, as in sunflowers or corn paired with pole beans or cucamelons. Using your vertical space wisely means planning for mature height, and understanding how and where the eventual shade will land and cool the surrounding soil.

Planting lower light vegetables under vertical structures capitalizes on the available light while the vining crops establish. This technique can be used for lettuce and herbs throughout the growing season, though dappled sunlight is ideal.

Proper Plant Spacing: Interplant, Don't Overplant

Proper plant spacing is crucial for productive gardens. Planting too close is all too common, as your eyes and ambitions exceed the limitations of your physical space. Ideal plant spacing allows for interplanting of two different maturing crops to succeed, creating multiple successions in the same space without compromising quality. Follow these guidelines for minimum plant spacing. If you can give your plants even more room between rows, that is another row of radishes, arugula, spinach, or leaf lettuce you could easily tuck between some young kale or broccoli.

When I space my brassicas 24 in. apart, there is always room for at least one succession of radishes in early spring. I also enjoy creating an "X" with five broccoli or mixed cabbages, flagging them in a diamond pattern with head lettuce that will crop several weeks before the broccoli matures. Spinach fits neatly between red cabbages, and despite its massive leaves, I harvest often; it bolts before the cabbages start to dominate the space. I have, however, tried beets with broccoli, and that resulted in shaded-out beets, as the broccoli and cabbage canopy closed faster than I'd hoped. Cauliflower grows more upright and works better with beets, and I imagine might also work for carrots as well.

As you explore interplanting, you will develop your own best practices and misfit pairs as well.

MINIMUM PLANT SPACING GUIDELINES

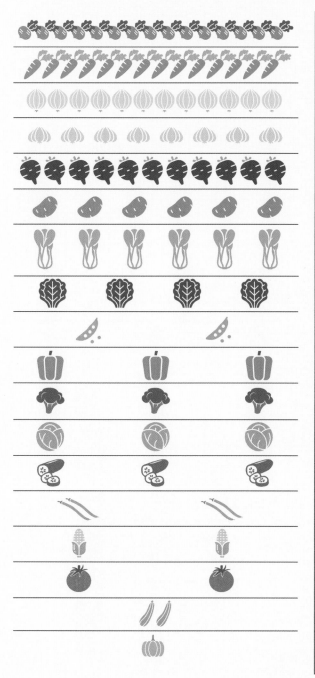

Radishes thin to 1 in. between plants, row spacing 4 in.

Carrots thin to 1–2 in. between plants, row spacing 10 in.

Onions plant 4 in. apart, row spacing 12 in.

Garlic plant cloves in fall 6 in. on center

Beets plant 4 in. apart, row spacing 12 in.
Kohlrabi plant 4 in. apart, row spacing 12 in.

Potato plant 9–12 in. apart, row spacing 24–30 in.

Bok choy plant 12–18 in. apart, row spacing 18 in.
Head lettuce plant 10–12 in. apart, row spacing 12 in.

Kale plant 12–18 in. apart, row spacing 18 in.

Snow and snap peas plant 1–2 in. apart, in double rows 3 in. apart, 24–30 in. between double rows

Bell pepper plant 18 in. apart, row spacing 24–30 in.
Hot pepper plant 12–18 in. apart, row spacing 24 in.

Broccoli plant 18 in. apart, row spacing 24 in.
Cauliflower plant 12–18 in. apart, row spacing 24 in.

Cabbage plant 12–18 in. apart, row spacing 18–36 in.
Brussels sprouts plant 18–24 in. apart, row spacing 30–36 in.

Cucumber plant 12–18 in. apart, row spacing 30–36 in.

Green beans thin to 2 in. apart, row spacing 24–30 in.

Corn thin to 6–7 in. apart, row spacing 30 in., minimum 4 rows for good pollination

Tomato plant 24–30 in. apart, row spacing 36 in. or 30 in. on center

Summer squash plant 36–48 in. on center

Winter squash plant 24 in., 36 in., or 48 in. apart in rows spaced 6–12 ft. apart for small, medium, or large squash

Interplanting Annual Flowers in the Food Garden

Flowers are the foundation of any successful garden, food or otherwise. Be it spring tree pollen essential for early native bees or sweet alyssum for predatory and beneficial hoverflies, flowers provide myriad benefits. Because so many summer favorites require pollination, interplanting vegetables with flowers in the garden is an integral and logical succession. Their colorful, playful layers bring visual interest, joy, insect diversity, and health to your food garden and local ecosystem.

The aforementioned best practices discussed above apply to interplanting flowers with food just as they apply to interplanting different food crops. You might have heard this called companion planting. Companion planting focuses more on the mutually beneficial relationship two plants create when growing together. I simply interplant with visual enjoyment in mind.

Edible flowers like these gem marigolds play perfectly into the hands of our pepper planting, massing together for a visual and edible feast.

Embrace the seasons

The longer you can grow flowers in your garden, the better—so plan for multiple successions of flowers just as you would your food. Some annual flowers like sweet alyssum arrive early and stick around throughout the growing season, while others, like zinnias, arrive as mid-successions in the height of summer, and are frost-tender. There are even late succession stars waiting to dazzle with pops of color and new life late in the growing season, like dahlias and sunflowers.

Our auxiliary perennial gardens are definitely the stars of my family's flower succession garden. Yet our food garden supplements the surrounding nectar buffet, bringing the pollinators even closer our prized vegetable garden. It's not just home gardeners who seize the benefits of flowers in their food garden: sage organic farmers plant annual flower gardens to attract beneficials and pollinators, space wisely used to promote health and diversity.

Location, location, location

Your garden real estate is precious. With food production the focus, select flowers as complementary elements. Edges are great places for flowers: along the edges of raised beds; underneath towering indeterminate tomatoes vines; at the base of a cucumber trellis; even among your compact summer companions like peppers and eggplants. Edges provide ample space outside the prized square footage of the bed for flowers to roam and flourish, adding whimsy and beauty while maintaining the core focus of food.

This sunflower planting in late summer is a haven for goldenrod soldier beetles and native bees in the middle of a 140-acre organic vegetable farm.

I always interplant flowers with every main season crop we grow. I have learned over the years who works best in our garden, and with our style, and I tend to lean into a utilitarian group for most of my interplantings. These flowers serve as the understory plants of our food garden. In other words, they harmonize with our main crops without dominating the space. They support from the sidelines, providing everything you could ask for in a flower growing among your food. They are so cheerful and functional that, when asked by garden visitors why I plant some of these, I often just let the plants speak for themselves.

Repetition for the win

Our eyes are drawn to repetition, a design best practice in formal landscapes. One aspect I love about bringing flowers into my food garden is the opportunity to unify the diversity of textures and foliage through flowers using repetition. Planting a few key annual flowers throughout my large garden unifies the landscape. And no matter the size, the presence and repetition of annual flowers across your food garden invokes a foundation on which your eyes can settle, as well as wander, and relish the edible beauty.

Natural insectaries

In addition to these flowers' beauty repeated across the garden, they do double duty as beneficial insectaries, attracting the good bugs to feast on the bad bugs. They are an indispensable succession in this regard, adding diversity every garden will benefit from. These are the types of beneficial plants a food garden should never be without—and I know these flowers will always be present in our garden.

Flower foundations

Some fantastic summer flowers can be added to a food garden as foundation plants, used in striking ways to visually and physically anchor the space. Sunflowers come to mind, towering in late summer, a beacon of hope for hungry bees. Sunflowers also happen to be emblematic of how you may feel as a gardener by late summer: more than a bit ragged, but smiling from ear to ear and soaking up all the sunshine before the season fades.

Zinnia are my other favorite foundation flower in my food garden. They are a generational nod to my great grandmother's garden, a place that, while I never knew it personally, lives on through my mother's stories of searching (often in vain) for the vegetable her grandmother requested her to retrieve. Zinnias remind her of those special seasons, and have remained an annual anchor in our family's gardens as long as I can remember. Carving out space in your food garden to honor your ancestors is its own kind of garden structure, nourishing your sense of joy, gratitude, and connectedness all season long.

ABOVE
Lacewing larvae are voracious predators of soft-bodied insects like aphids and scale. A sign of a healthy garden ecosystem, this green lacewing larva was spotted in late spring enjoying annual self-seeding chamomile in our herb garden.

RIGHT
Self-seeded sunflowers anchor this late summer garden while the hummingbirds drop in to sip raspberry flowers. A cover crop of buckwheat in full bloom grows adjacent to the raspberries.

Garden generalists

One of my preferred interplanting strategies employs low-growing flowers that reach heights of less than 24 in. I consider these annual flowers garden generalists. Many thrive at a wide range of light levels, can handle the shoulder seasons, and spread horizontally, rather than vertically, becoming a luscious green mulch.

My favorite generalists are sweet alyssum, marigolds, calendula, and nasturtiums. All these flowers are edible, so they are naturally right at home in the food garden. Because of their low profile, they make suitable companions between nearly any vegetable. While I've developed a personal aesthetic for planting them, it is merely one of countless ways to interplant flowers with food.

I love a long line of nasturtiums in my orchard; the variegated foliage of Alaska is one I'm rather partial to. A massing of dainty white alyssum under my tomatoes is pretty much non-negotiable in our food garden. I relish the scent of the medicinal herb calendula as I brush past my eggplants and cabbages. The citrus fragrance of gem marigolds will always remind me of my pepper garden, because that's where I interplant most of them.

PAIRING LOW-GROWING FLOWERS AND VEGETABLES

Flower	Pairing Tips
Sweet alyssum	Plant with tomatoes for green mulch; beloved by hoverflies
	Repeat planting at bed edges for harmonious statement
	Plant with herbs, strawberries, brassicas
Marigold	Plant with peppers, tomatoes, cucumbers, squash, beans
Calendula	Plant with beans, brassicas, herbs, eggplants
	Flower early and late season; slower to flower in height of summer
Nasturtium	Massive tufts are excellent bed border
	Plant under vertical squash trellises
	Plant under fruit trees

Rosie O'Day sweet alyssum effortlessly flops itself into the paths, creating a lush summer garden view.

OPPOSITE
Growing up is one of the most efficient space savers for urban gardens. If it vines, you can trellis it.

I lean into the wisdom of prairies when I plan and plant my annual flowers. The prairie's biggest success is how it provides refuge and food for insects and other creatures in all four seasons. Its continuous stream of blossoms in the growing season is never without a succession or two in peak "ripeness." Its rotating menu of nectar and pollen is where I draw inspiration when planning and planting flowers in my food garden. While marigolds and nasturtiums are main season successions, calendula and sweet alyssum shine in both the spring and autumn shoulder seasons, making these four flowers an indispensable grouping.

More than anything, remember that interplanting boosts your food garden's productivity across the growing season immeasurably. However you choose to implement these strategies, find the approach to that suits your gardening style, local climate, and site requirements, and embrace this philosophy. Interplanting is guaranteed to propel your garden forward in leaps and bounds.

Small Space Strategies

Succession gardening is a tool for every garden, but especially small spaces. Urban and container gardens are where this method can truly transform your growing experience. Your relationship with your containers, single raised bed, or balcony oasis is undoubtedly one of great joy. As always, succession gardening can deepen that joy by providing more food and flowers for longer.

Prioritize vertical structures to maximize your garden. Indeterminate plants are often vining and unruly, and vertical structures provide the extra space for those plants to thrive while creating microclimates below, where unlikely season-pushing magic can happen. Some late radishes might dwell in the shade of a leaning cucumber trellis; summer lettuce may linger longer thanks to the cool of the shade. Flowers eagerly flourish, cascading out of a container or in front of a bed. Bush beans tumble out of the front of a container while behind them carrots, basil, and tomatoes rule the roost.

Pushing the limits of your growing season is critical in small spaces, even more so than larger succession gardens. If you start extra early, you harvest that much more food. On the other side of the growing season, try to establish cool season crops before the mercury dips.

While flowers are indispensable, if I my space was limited, I would look for more space adjacent to my food garden to add flowers. That might mean adding another container, hanging basket, or converting an adjacent portion of lawn to a cut flower garden so your sunniest and most productive space is focused on food production.

Summer gaps
are renewed
with fall plant-
ings, both direct
sowed and trans-
planted, creating
a multi-succession
food garden by the
middle of August.

GARDEN RENEWAL: SUCCESSION PLANTING IN SUMMER

The traditional vegetable garden is finished being planted by summer. Meanwhile, the succession garden has already fed you well, and while close to fully planted, it is about to be planted again. Summer is a key moment of the growing season for succession planting. Time is on your side, and so are growing conditions. Your focus now should be anticipating mature crops and their imminent gaps in the next four to six weeks after summer solstice. Northern climates have a narrower window in which to replant than warmer climates, but everyone has the opportunity. That opportunity will keep you anticipating and renewing the garden in earnest with as many successions as possible all season long.

Perennials round out the fall harvests, our first apples blending with a summer and fall vegetable medley.

The most creative aspect of the garden plan is when and what you replant your garden with in the height of summer. When the opportunity arises to replace a bed, I'm ready with more fall brassica seedlings, or a short list of mid-season crops that are excellent direct-seeded successions, such as sweet corn, carrots, beets, and bush beans. As my warm season summer crops inevitably fade, my fastest brassicas come back to close the growing season with an abundance of quick, cold-hardy foods that hold well in the chilly autumn weeks, effortlessly extending our harvests.

Commit to Renewal of the Garden

Earlier in my gardening journey, I was committed to my main plantings once summer set in, marveling at the constancy of the chosen plants that adorned our space. It was the right garden for me at that time, providing plenty of food in an urban setting with room to grow. But I'd like to say I've grown wiser over the last decade or so. Life is not meant to go on forever, and neither is the garden succession that peaks in summer.

I can now see that the garden's inevitable senescence is not the end of the season. Rather, it is the beginning of a new season within a season. These are prime moments for new life to spring forth, just as the forest makes haste to repopulate a gap. It is our opportunity as succession gardeners to anticipate and be prepared to fill these gaps. Where I once lamented the end of a crop's life cycle if it occurred before first fall frost, I now see that end as an opportunity to extend the season and increase productivity.

Sometimes disease comes along and takes a tree or plant. Again, I look to our forests and see how a diverse ecosystem weathers disease more robustly than a site planted with a single species. While we have lost several fruit trees to extreme winter cold in our brief time in our current garden, we have also gleaned our first fruit from a few of the trees that remained healthy. As we replace the lost trees, it sets us up for an extended maturation of our orchard. Our replacement apple trees may mature when our young pears finally fruit, for example. Their journey and ours will carry on.

JUNE SOWINGS

Indoor Sowing	Direct Seeding
Celery Heat tolerant lettuce and bok choy Broccoli Cabbage Cauliflower Kale	Carrots Beets Green beans Sweet corn Dry beans Summer squash Cucumbers Winter squash Melons
Indoor or Direct Sow	
Herbs, especially basil and dill	
Flowers, especially branching sunflowers	

The Path to Renewal in Summer

There are several distinct ways to find space in a summer garden to plant a new succession. Some are intentional and planned, others opportunistic. First is the natural progression of a planting as it reaches maturity. Any spring crop that completes its life cycle opens a natural gap in the garden, likely by or during July. These spaces open up earlier when you start your season as early as possible, but either way they mature with plenty of time for a second succession.

If you've carefully cultivated your seed collection to include varieties with varying days to maturity for staggered harvests, you already have a steady stream of succession for the coming weeks and months. This alone is the simplest succession garden plan: plant multiple types of every vegetable, all cropping at slightly different times, thus providing a steady stream of sustenance without completely inundating your countertops. . . . Well, I can't promise that last part, but it's definitely the goal to grow a garden that produces for longer, though at a rate commensurate with my stamina and ability to efficiently process and enjoy it all.

Garlic that matures in late July naturally opens a gap in the middle of the growing season, a key opportunity for maximizing food production and capitalizing on fall plantings.

The Revolving Door

Your spring peas fizzle when summer's heat kicks in. Spring cabbages and broccoli are harvested, and space opens for summer sowings of beets, carrots, green beans, or heat tolerant lettuce. Your bush dry beans mature six weeks before first frost; you make haste and remove them, and gain square footage for a fall garden. Your succession garden plans take into account the maturation dates of all your crops, across the seasons. With an eye to those vegetables that swiftly fill the gaps to keep the harvests steady, you retain a planting mindset all season long.

CHOOSE YOUR NEXT PLANTING:
HOW MANY WEEKS UNTIL FIRST FROST?

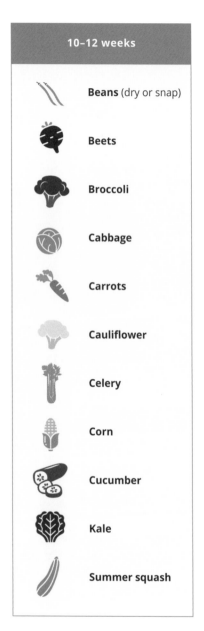

10–12 weeks

- **Beans** (dry or snap)
- Beets
- Broccoli
- Cabbage
- Carrots
- Cauliflower
- Celery
- Corn
- Cucumber
- Kale
- Summer squash

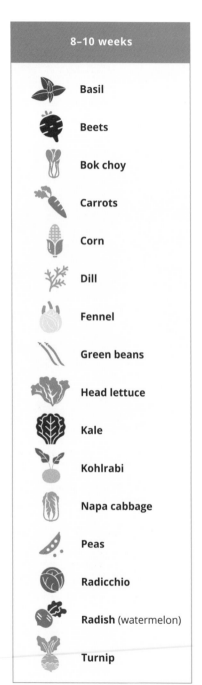

8–10 weeks

- Basil
- Beets
- Bok choy
- Carrots
- Corn
- Dill
- Fennel
- Green beans
- Head lettuce
- Kale
- Kohlrabi
- Napa cabbage
- Peas
- Radicchio
- **Radish** (watermelon)
- Turnip

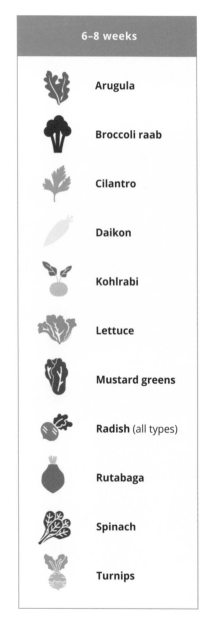

6–8 weeks

- Arugula
- Broccoli raab
- Cilantro
- Daikon
- Kohlrabi
- Lettuce
- Mustard greens
- **Radish** (all types)
- Rutabaga
- Spinach
- Turnips

SUMMER GARDEN TASKS

Seed starting	Sow cool season crops indoors as fall garden transplants
	Continually assess opportunities to renew or add successions; adjust seed starting accordingly
Direct sowing	Try carrots, beets, string beans, dry beans, and corn as second successions
	Direct sow warm season crops a second time, including cucumbers, melons, and summer squash
Pest management	Begin to look for common garden pests and proactively remove them
Bed maintenance	Between plantings, amend soil with slow-release organic fertilizer when seeding or transplanting
	Edge and weed beds regularly
Perennials	After strawberry harvest, renovate bed to encourage strong plants
	Lightly prune fruit tree suckers
	Continue pest management
	Apply kaolin clay to fruit trees as physical barrier against pest damage like Japanese beetles and plum curculio

Growing Across the Seasons

Dipping into my experiential encyclopedia of what vegetables need to thrive, I work to develop and implement a sound yet flexible plan for our garden. This includes prioritizing our favorite vegetables (brassicas); playing with the edges of our seasons; being ready with hardened off vegetable starts all season long; and fully embracing the sometimes chaotic pace of the succession garden. The beauty is that every single day is an opportunity to enter into that garden plan, for there is always something that would be appropriate to plant, either by direct seeding or as indoor starts.

SUMMER SUCCESSION PLANTING GUIDE
FOR EXTENDED FALL HARVESTS

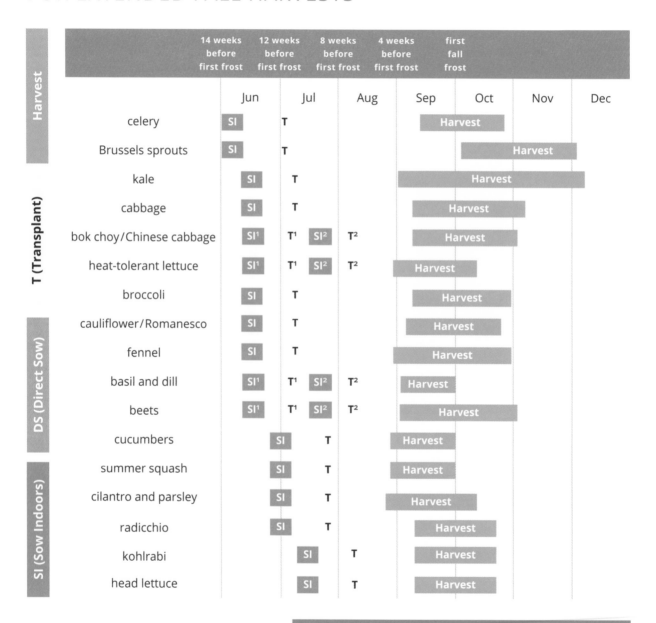

	14 weeks before first frost	12 weeks before first frost	8 weeks before first frost	4 weeks before first frost	first fall frost			
	Jun	Jul		Aug	Sep	Oct	Nov	Dec

Harvest

T (Transplant)

celery	SI	T			Harvest			
Brussels sprouts	SI	T					Harvest	
kale	SI	T			Harvest			
cabbage	SI	T			Harvest			
bok choy/Chinese cabbage	SI[1]	T[1]	SI[2]	T[2]	Harvest			
heat-tolerant lettuce	SI[1]	T[1]	SI[2]	T[2]	Harvest			
broccoli	SI	T			Harvest			
cauliflower/Romanesco	SI	T			Harvest			
fennel	SI	T			Harvest			

DS (Direct Sow)

basil and dill	SI[1]	T[1]	SI[2]	T[2]	Harvest			
beets	SI[1]	T[1]	SI[2]	T[2]	Harvest			
cucumbers	SI	T			Harvest			
summer squash	SI	T			Harvest			

SI (Sow Indoors)

cilantro and parsley	SI	T			Harvest			
radicchio	SI	T			Harvest			
kohlrabi		SI	T		Harvest			
head lettuce		SI	T		Harvest			

extend harvests by protecting mature veggies with frost blanket

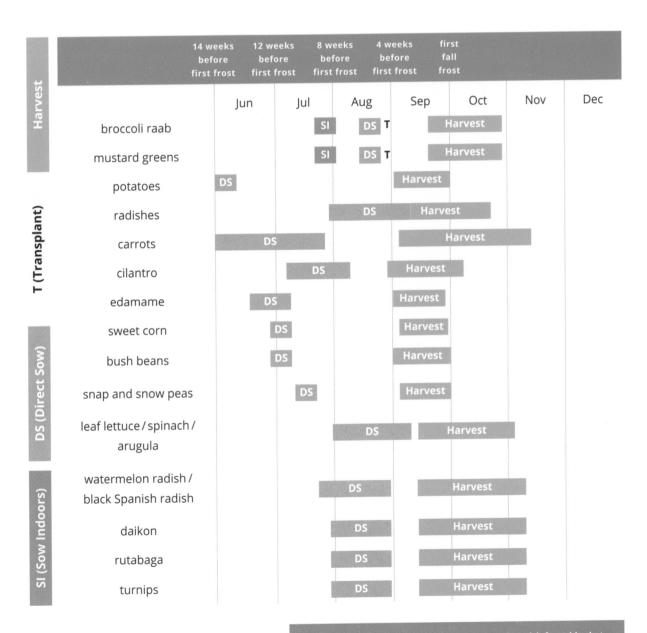

	14 weeks before first frost	12 weeks before first frost	8 weeks before first frost	4 weeks before first frost	first fall frost		
	Jun	Jul	Aug	Sep	Oct	Nov	Dec
broccoli raab		SI	DS T	Harvest			
mustard greens		SI	DS T	Harvest			
potatoes	DS			Harvest			
radishes			DS	Harvest			
carrots	DS			Harvest			
cilantro		DS		Harvest			
edamame		DS		Harvest			
sweet corn		DS		Harvest			
bush beans		DS		Harvest			
snap and snow peas		DS		Harvest			
leaf lettuce / spinach / arugula			DS	Harvest			
watermelon radish / black Spanish radish			DS	Harvest			
daikon			DS	Harvest			
rutabaga			DS	Harvest			
turnips			DS	Harvest			

extend harvests by protecting mature veggies with frost blanket

Harvest

T (Transplant)

DS (Direct Sow)

SI (Sow Indoors)

These plans are most successful and productive when they include as much diversity as possible. Diversity not only provides insurance against inevitable pest and environmental loss, it avoids the seasonal overabundance that is a hallmark of the traditional summer garden. While a dining table full of tomatoes and summer squash is an awesome sight and a late summer tradition, it is also a literal and proverbial late summer weight. Succession gardening takes that weight and disperses it more evenly across the seasons.

Summer offers a broad window of opportunity for replanting. During these flexible weeks, resetting beds provides an endless field of possibility. As you approach first fall frost, your list narrows, and you should lean into the fastest-maturing crops as your most reliable options for replanting. As personal as your garden is to you, the how, when, and what of removing and replanting is uniquely your decision. This has become a fluid process in our garden, accepting the idea of early removal in favor of a new crop. The more I embrace this turning over, the more that embrace renews my energy and excitement for tending to a wild late summer garden.

A fall garden in the making, late season brassicas are added to the gap created when I took out the pickling cucumbers and sweet corn.

Example Succession Garden Layout

Each row of this table is a specific bed, each column a succession. As you follow the successions across each row, you are working your way across our growing season, seizing opportunities to replant, renew, and extend your harvests.

First Succession	Second Succession	Third Succession
Early cabbages, interplanted with radishes under row cover, harvested early June	Direct seeded bush beans, interplanted when cabbages near maturity	Late August renewal with Asian greens and radishes
Early spring beets, harvested June through July	Direct sowed fall carrots, bush beans, late sweet corn, edamame, or transplants of fall cabbage starts	Beans and sweet corn can be followed by fastest early fall crops: arugula, radishes, leaf lettuce, broccoli raab transplants, and spinach
Garlic, planted in fall, harvested mid- to late summer	Fall brassicas, interplanted with lettuce and beets	None
Early peas, interplanted with radishes, lettuce, and carrots, fully harvested by mid-July	Mid- and long season cabbages, transplanted mid-July, harvested September through November *or* Second succession of summer squash transplants	None
Spring kohlrabi and bok choy, planted late March under cover, harvested by late May	Direct sowed main season carrots, harvested late summer through late fall *or* Second sowing of bush beans, beets	None
Onions, planted from starts early May, harvested late July to early August	Fall brassica starts, sowed July, planted August, interplanted with direct-seeded radishes *or* Fall peas (experimental, highly dependent on the late summer climate)	None

First Succession	Second Succession	Third Succession
Buckwheat cover crop	Fall cabbages, cauliflower, celery, and broccoli, interplanted with heat tolerant lettuce	None
Red cabbage interplanted with various head lettuce	Direct sowed fall carrots or beets; transplanted fall celery or heat tolerant head lettuce; direct sowed late bush beans	None
Summer squash, first planting (removed in late July)	Direct sowed fall root crops; examples include watermelon radish or daikon	None
Pickling cucumbers, finished mid-August after canning or due to disease	Direct sowed mustard greens, broccoli raab, mizuna, kohlrabi, other brassicas maturing under 50 days *or* Transplanted Chinese cabbage starts *or* Direct sowed fall roots: turnips, daikon, watermelon radish, globe radish	None
Sweet corn, planted early May, harvested early August	Any fast-maturing fall brassica, such as kohlrabi, mustard greens, broccoli raab, bok choy, Chinese cabbage	None
Early determinate tomatoes, removed from garden two to three weeks before first frost	Radishes, arugula, spinach, or garlic (fall planted for next year)	None
Early potatoes	Second succession of summer squash (8 weeks before first frost) *or* Direct sowed late succession of bush beans *or* Direct sowed fall root veggies	None

Some vegetables are simply space and time hogs. They also happen to be some of the most well-loved, productive foods to grow and eat—tomatoes, squash, beans, and potatoes come to mind. Incidentally, they produce heavily, likely the very reason we collectively lean into them as home gardeners. Generally, I don't subscribe to intensive interplanting with these crops, but instead provide them with ample space, nutrition, and supplemental irrigation so they will feed us well all summer—and with proper storage measures, well into winter too.

Besting Pest Pressure

Knowing your pest pressure and having a plan for mitigating it is key in garden planning. Rabbits once feasted on our urban blueberry shrubs in winter, chewing them down to the snowline one year. We retaliated the following autumn with custom-built chicken wire frames. Small rodents, birds, and other mammals, not to mention insects, will all visit your garden at some point. This is where diversity will play a role in your succession garden planning.

Intense bird and rodent populations reside all around our food garden, but we haven't yet lost a single tomato or apple due to wildlife pressure. So far, our blueberries are only enjoyed by humans. Our fence is our first line of defense, but it does not protect against rodents and birds who freely come and go despite my maniacal outbursts and colorful hand gestures. While I can't unequivocally explain our good fortune, it is highly plausible that our auxiliary gardens—including the likes of viburnum, elderberry, and the multitude of flower seeds dotting the prairies—provide enough to keep them satisfied and well fed.

Pests come and go year to year. This cabbage looper arrived suddenly a few years ago, generalists of much more than just brassicas, and a master of camouflage.

Trap crops are another common method for besting pest pressure. Trap crops are placed some distance away from your food crop, and planted ahead of your desired crop. In our climate, planting squash early to entice squash vine borers and removing their eggs is feasible, because we only have one life cycle per year. Floating row cover is another form of pest management common to organic gardens. Personally, I've found I only trade off other pest issues

under the row cover, as it gives me a false sense of security. I prefer to keep a closer eye on my plants and forego row cover, and remove insects by hand.

While we try to stick to compost and fertilizer as the only inputs for our garden soil, there are instances that necessitate more. Beneficial nematodes are microscopic predators that feed on a whole host of soil pests, and we have used them in our garden several seasons in a row to combat root maggots, Japanese beetles, and flea beetles. There are specific species that are active at different soil depths, thus more apt to consume different species of pests. This approach requires a fair bit of financial investment, but we've found it has had a lasting positive effect on our soil health. I feel it paid for itself in the quality of produce and peace of mind we reaped as a result.

Your best defense and offense against pests in the organic garden is your time. Take the time to monitor your plants. Use your hands and eyes to inspect and manually remove larvae. Get on your hands and knees and really eyeball your leaves and stems. Get to know your garden insects, and identify insects you discover before making any decisions about their fate—unless you know they are only going to skeletonize your bean plants all summer, as is the case with Japanese beetles. More often than not, the insects you find in your garden serve a purpose, are an integral part of the local food chain, and won't cause severe crop damage in your garden.

Our hands are the best weapons against Japanese beetles, and it's a daily chore in morning and evening to plunk them into a pail of soapy water. They gravitate to our corn silks, pole beans, edamame, raspberries, and tend to congregate (if you catch my drift) in the asparagus patch too.

Rolling with the Changes:
Pest Pressure as Succession Opportunity

Every growing season inevitably carries with it a new disease or pest. Not only does this yield new knowledge, it also offers the opportunity to reset your garden with the next best thing. Unanticipated openings in the garden are

heartbreaking, as when you have to pull diseased plants in the middle of the growing season. But after tending your bruised ego, take a good look, consider what's growing, what's about to crop, and what new seeds would mature in a few months' time. Fifty days is the shortest amount of time I could feasibly harvest a warm season crop when summer heat is turned up. Cucumbers, bush beans, and summer squash come to mind as great succession plantings for unplanned gaps when your first fall frost is at least two months away.

One summer, I prematurely harvested a troubled Yukon Gold potato planting, in spite of my desire to hold onto the declining plants. I transformed the space into a second succession of summer squash and a first sowing of fall root vegetables. I could have lamented their premature departure, tried to pamper and just hold on to them a little longer. Instead, I set them free, accepting their signals that they just weren't having it. Yukon Gold is known to be a finicky potato, and while we don't generally harvest potatoes until September (for long-term storage), we picked and enjoyed these through the remainder of the summer instead. Meanwhile, our other potato varieties remained vigorous and healthy, ready to provide for winter meals. This unanticipated gap was a welcome twist that made summer meals a little more interesting, and reminded me to stay nimble with all my crops.

How you negotiate plantings that end up less than ideal is completely up to you. For me, the cathartic reset of a hot summer crop in favor of a quick fall succession is a proactive way to manage the garden's late summer abundance.

Sometimes a garden bed isn't living up to expectations. Perhaps you aren't enjoying the produce as much as you'd hoped, or maybe it's not as vigorous as you expected. I am here to tell you it's more than okay to remove those unproductive or unpalatable plants in favor of a new planting. It may seem ruthless to pull out flowering pole beans, but when you aren't in love with their flavor and their productivity is pitiful, it's time. Why not reset with something that will bring more joy to your plate? I would lean into something trusted, tried, and true. Succession gardening is especially conducive to creative flexibility in service of these moments.

At what point do you say "Enough!" to such a planting? My last call for resetting unproductive or diseased beds is six weeks before my average first fall

Telltale of anthracnose fungal disease are these concentric rings and the subsequent bullseye decay of the leaf; the lesions eventually cause foliage to fall off.

OPPOSITE
After pulling out beans that did not impress, knowing we were six weeks from my first frost, I leaned into my hardened off summer starts and renewed the unproductive planting with beets, bok choy, head lettuce, and kohlrabi.

frost. Lately, I scour the late summer garden with a deeper sense of urgency and opportunity for volunteers that may be ready to head to the compost bin prematurely. Leaning into the deep library of cold-hardy brassicas, I have plenty of other plants to sow in late summer and enjoy throughout autumn.

The other important time to contemplate turning over a bed is when your plants show signs of disease, even if they are just getting going. This is not a mandate for all plants, as there are plenty of diseases I simply cut out and work around; yet I have learned that those affected by some fungal diseases simply aren't worth holding onto for the sake of a few more fruits.

One example that comes to mind is a hot, dry summer unlike any I've known in our climate. One side effect of drought was increased fungal pressure, something I only understood in hindsight, likely exacerbated by my lackadaisical supplemental irrigation. We lost cucumbers and melons to a severe case of anthracnose, and removed both in the middle of what should have been their prime. It was a dagger to my gardening heart, and I did my best to receive it humbly, with open arms, as a growth opportunity.

This garden renewal is an act of becoming. Every year, new ways of renewing the garden offer themselves to us. Seed starting throughout the growing season provides the foundation for these opportunities. I sow trays in spite of my fully planted garden, and often lament the lack of space for these future seedlings. But there's another part of me that quietly recognizes the shifting landscape, anticipating imminent gaps, and trusting it will all work out.

SUMMER SOWINGS FOR THE FALL GARDEN

Indoor Sowing	Direct Seeding
Broccoli Cabbage Bok choy Chinese cabbage Kale Cauliflower	Watermelon radish Daikon Rutabaga Carrots Cucumbers (early July) Green beans (bush) Peas/edamame Dill (herb) Mustard greens

Indoor or Direct Sow
Heat-tolerant head lettuce
Kohlrabi
Green onions
Beets
Calendula
Cilantro
Sweet alyssum (June)
Sunflowers

The Wilding of August

You know the moment: your tidy, well-behaved, youthful garden suddenly becomes an unruly teenager. It happens slowly, so it takes you a bit by surprise. You tell yourself this year will be different. But it never is. Once innocuous squash vines completely engulf the paths. Dozens of split cherry tomatoes dot the garden bed, dispersing seeds, thus securing next summer's weeding project. Sunflowers tower over the garden, perfectly performing your inner soundtrack: ragged and tired, but still here.

Late summer gardens overwhelm with abundance, and there's no single path to gracefully carrying on. Instead, I follow my heart, which means I vacillate feverishly between soaking up all the chaos, reminding myself of the barren winter landscape looming on the horizon; and complete despair over the accumulated biomass and endless work that lies ahead.

The expansion that began in early June has fully arrived, flourished, and provided sustenance and joy. Now my body and heart are beginning to turn inward, ready to gather the big harvests and reflections on the year. By the end of August, I am ready to rip out my diseased tomatoes, even without a plan for replanting. I am just plain ready to deconstruct the annual garden, getting ahead of the fall cleanup. And sometimes that's just what we do. Clearing the space that no longer serves us this time of year is its own very specific renewal, literally and figuratively.

Layering Successions in Summer

Warm season vegetables establish rapidly when sown at the right time. Too soon and you'll be met with sluggish germination and a predisposition to fungal pressures like damping off disease. Done right, however, warm season vegetables play perfectly in the succession garden plan.

Consider sowing more midseason successions. Summer squash planted early July will produce well in late summer, and be tamer and tidier than your first squash plantings. Keep sowing green beans, especially fast-maturing and space-saving bush types, until eight to ten weeks before first frost. If you missed sowing cucumbers in June, simply sow by early July for a late summer harvest. This is the season of flexibility in the home garden, and the more you play, the more diverse, manageable, and extended your harvests will be.

The beauty of summer is how the bounty of warm season crops mingles with the resilience of the cool season producers. Summer offers weeks upon weeks of succession planting options. Certainly one or more of your favorite vegetables can be succession planted, both to invite earlier harvests and extend further into the late season. Taking the time to plant food that will produce these staggered harvests, as well as continually adding crops with varying maturation rates, is guaranteed to generate a rich and delectable tapestry of food in summertime.

Our harvests are diverse throughout the growing season because we sow seeds steadily from February's peppers and onions all the way through to August's tidy rows of radishes.

OPPOSITE
The chaos of August: love it or hate it, the best route is to try to sit and be present with it.

Succession planting cold-hardy vegetables throughout summer provides steady, bountiful, and diverse harvests well beyond our first killing frost.

THE FALL GARDEN AND EXTENDING THE HARVEST

In addition to getting ahead in spring and keeping up in summer, the fall garden is equally vital to any succession plan. Autumn is the season of harvesting and gathering, from literal baskets of food to proverbial garden and life lessons. This season is steeped in the knowledge and appreciation of accumulated growth, both personal and literal. By the time autumn fully settles in, both you and the garden are a bit weary, and as easy as it is to shut the gate until next year, autumn is the ideal season for reflecting while the lessons are fresh. Like spring, it is a season filled with vegetables that thrive in the margins. These foods grow when others simply cannot, making stellar garden companions in any climate. Without being intentional about the fall garden, our season would abruptly come to a halt by early October, with kale, leeks, carrots, and Brussels sprouts taking the final bow.

Almost all the vegetables and flowers in this view will shake off the light frosts that are about to arrive, and many taste better after a few good cold snaps concentrate sugar in their roots.

OPPOSITE TOP
Fall harvests reflect the muted tones that are slowly taking root all around us.

OPPOSITE BOTTOM
If at first you don't succeed, try, try again. Bush beans made for happy early autumn meals in this year, as our earlier successions did not produce as splendidly.

The fall garden is a game of mental stamina to be sure. As you start summer tomatoes and peppers in late winter, so too must you sow parts of the fall garden in late spring. Many of these garden successions takes a full season to grow and mature. I sow our earliest, slowest to mature fall cabbages at the same time I transplant my first cucumber and squash starts into June's warm, welcoming soil. Even though late spring might seem much too early to sow these crops, we are already fourteen weeks from first fall frost (around 100 days) at that point, and that is precisely the right time to start that fall garden.

Fall gardening means embracing flavors you normally might not, for those are the flavors of the sturdiest and hardiest foods for these otherwise inhospitable weeks. Some vegetables (root vegetables in particular) I grow only in autumn, because I prefer their flavor after a few frosts sweeten their roots. My favorite fall root vegetables include beets, carrots, daikon, watermelon radishes, and turnips. And each climate offers its own sweet spots for fall vegetables. The most important thing is to begin sowing with enough time to reach maturity before consistent hard frosts.

There is a wide range of heat tolerance and days to maturity among fall garden crops. Most of our fall cabbages are seeded in early June and transplanted early to mid-July, forced to settle into the garden during our first 95°F (35°C) heat wave. Because most require 80 to 90 days to reach maturity, this is the appropriate time to transplant our primary fall brassicas.

As the heat subsides in August, the plants are well established. As the light fades in September, they are in various stages of maturity. By the time October arrives, they are harvestable and simply sit in what's become our outdoor refrigevrator until needed. Before consistently hard frosts settle in, we harvest our mature brassicas and store them in our root cellar.

It takes some trial and error—and mental fortitude—to find your path to success with these sturdy vegetables. Many a year I have planted either too early or too late. Mature broccoli rolling in by late August, during the height of summer harvests, is not a fall garden; softball-sized cabbages in December also misses the mark. How I compensate for inevitable variability during these months is by sowing multiple successions, all with varying days to maturity and heat tolerances. Our fall harvests reliably commence by September as the tomatoes fade, and we welcome back into the kitchen many of our beloved culinary companions.

In addition to the reliable brassicas we lean on when temperatures decline rapidly, we turn to other, less obvious early fall companions, including summer sowings of mid-succession crops like sweet corn and bush (snap) beans. We have grown corn both as early and as late as possible, as both a first and last succession, depending on available space. It works well both ways, and homegrown corn for Labor Day weekend is a bit of a novelty; it is equally exciting to aim for homegrown sweet corn by the Fourth of July. Bush beans are also a great way to extend your bean season. They take up little space, and if given the latter part of summer to establish, come into season as the main summer successions fade.

First hard freezes are bittersweet and yet marvelous. Delightfully geometric ice crystals signal the official shift to cooler days and hardier, tidier foods.

How to Get Ahead in Spring

Fall happens to be the ideal time to plan your future garden. We spent our first summer in our current home dreaming and scheming. Having relocated during the height of the growing season, we left behind an urban garden filled with summer squash, zinnia, cucumber, green beans, onions, peppers, tomatoes, and what would have been our first peaches. We spent our first months studying, sun watching, drawing, exploring, and generally getting the lay of the land. Gaining an understanding of the site's challenges, we focused our efforts on designing a large garden on a hill. We removed trees, erected fences, delineated beds, and amended soil. We were blessed with an extra-long autumn and nearly managed to complete garden prep before the first snow. We entered winter confident of a promising growing season come spring. All that getting ahead, breaking gardening ground in autumn, made a huge impact on the success of our first full growing season.

We made this a tradition after that first pre-season of gardening. Our daily garden dialogue in the fall may include talking through changes we envision, how to incorporate what we've learned, and what we're dreaming for next year's plans. We identify which beds will be our earliest plantings next spring, and prioritize them for late winter low tunnels. Our punch lists ebb and flow, but there are tasks we chip away at consistently each year between September and December.

FALL GARDEN TASKS

Planning	Build new raised beds, trellises, fences
	Test soil and amend as needed
	Develop system of recording successes and failures for future years
	Reflect and set goals for next year to sharpen focus
	Plan crop rotations using photos
Procuring	For best selection, order seed as early as your grow list is final
	Check seed stash; replace low quantities to keep on hand for succession planting
	Order ingredients for bulk fertilizer and mix
	Order bare root trees and shrubs, and pre-dig holes to plan for their spring arrival
Planting	Plant garlic in compost-amended soil after first killing frost, several weeks before ground freezes
	Transplant and mulch bunching onions sown in summer to overwinter for early spring harvest, even in the coldest climates
	Plant other bulbs, including deer-resistant daffodils; spring and fall crocus; or other spring and summer delights
	Sow extra-hardy greens around fall equinox for late season and early winter harvests under row cover, as in spinach and mâché
Harvesting	Harvest and cure shelf-stable foods before hard frost: winter squash, dry beans, flint and popping corn, potatoes, late season apples
	Harvest root vegetables after first hard frost and place in root cellar for extended storage: carrots, beets, watermelon radish, rutabaga, daikon, kohlrabi, etc.
	Keep fall brassicas in ground as long as possible before harvesting to sweeten and extend season
	Wait until late fall to harvest Brussels sprouts, leeks, and kale last, closing the harvest season
Removal	Remove all diseased plant material (including rotting fruit) to minimize diseases persisting in soil
	Dispose of diseased plants in municipal trash (unless homemade compost pile heats to 150°F (65°C) for a minimum of ten days)
	Add healthy garden plant material to fresh compost pile
Perennial maintenance	Identify and prepare area for new plantings
	Amend pH of blueberries as needed
	Thin strawberries to one young plant every 12–18 in.
	Cut down and remove all asparagus foliage after dieback to prevent asparagus beetle larvae from overwintering
	Protect young fruit tree bark with tree guards to minimize potential sunscald or girdling
	Again, add healthy plant material to fresh compost pile
Season extending	Top dress soil for earliest spring plantings with fresh compost
	Set up hoops for low tunnels in beds where earliest transplants will go
	Clean and organize your poly or row cover for winter use

Fall is the perfect season to make gains for next year. It's an ideal time both to establish a food garden and prepare for perennial plantings such as blueberries, strawberries, and fruit trees, most commonly planted bareroot in spring. It is also prime time to build raised beds, fences, and trellises. Any groundbreaking tasks that might delay the onset of your growing season come spring should be completed at the end of your current growing season. Consider this your first succession of the following growing season.

In a well-planted succession garden, the first freeze is a transition, but far from the end of the growing season. True, your garden will transform significantly as warm season crops fade and space opens up, however much will remain even beyond the first freeze. It's a homecoming, a return to the low-lying, cold-tolerant rows of sturdy brassicas and root vegetables of spring—a return to the familiar and quieter gardening season.

We plant our garlic two weeks after the first hard frost, sometime in mid- to late October. It's a task we generally put off until snow or cold command our attention.

Tasks leading up to the first fall freeze include harvesting frost-sensitive full season crops like winter squash.

FAR RIGHT
The colors of fall may be subdued aboveground, but bursts of pastels and vibrant hues of orange and red lay below, in the colorful and nutritious root vegetables that humbly await their harvest day.

Having bid adieu to indeterminate vining crops, the garden retains only those frost-tolerant and shelf-stable vegetables. Carrots stay put indefinitely; we harvest and store right before our ground starts to freeze. Kale, Brussels sprouts, cabbage, and many others are remarkably cold tolerant. Sowed in May, June, and July, these seeds are now the stars, having weathered 100°F (38°C) heat indexes for several weeks during their infancy. It is as awe-inspiring as the journey of a tomato seed.

Autumn is where the garden ultimately begins for me. Death is a renewal of life, an opportunity for rebirth. For some climates, this death-rebirth cycle occurs continuously all year. For other regions, the height of the summer is the dormant season. But for most, it's the first freeze that transitions the garden from active to passive.

More commonly called first frost, the first fall freeze marks the end of most home garden plantings. While these terms are used somewhat interchangeably, they are two distinct meteorological events. Not all freezes result in frost, and frost—that is, visible ice crystals—can form without a freeze. It's a function of the dew point when the air masses drop below freezing, so dew points will determine whether a frost occurs. Air temperatures are the measurement of a freeze, and it's the freeze that truncates the season.

Cold Tolerance of Common Succession Plants

In my experience, many of these plants withstand even colder temperatures than those listed here. These are temperatures to set your mind at ease and encourage you to push the season. I've seen kohlrabi, Brussels sprouts, and broccoli weather a low of 15°F (–9°C) unprotected without major frost damage. The question is how long you intend to store after harvest, or whether the crop is for immediate consumption. If long-term storage isn't your goal, then pushing your low temperatures even further should be met with curiosity and joy.

Plant	Freezing Temperature Threshold
Beets	28°F (–2°C)
Carrots	28°F (–2°C)
Radishes	28°F (–2°C)
Turnips and rutabaga	28°F (–2°C)
Peas	28°F (–2°C)
Sweet alyssum	28°F (–2°C)
Calendula	28°F (–2°C)
Viola	28°F (–2°C)
Snapdragon	28°F (–2°C)
Head lettuce	26°F (–3°C)
Cauliflower	26°F (–3°C)
Mustard greens	24–28°F (–4 to –2°C)
Kohlrabi	24–28°F (–4 to –2°C)
Bok choy	24–28°F (–4 to –2°C)
Broccoli	24–28°F (–4 to –2°C)
Cabbage	24–28°F (–4 to –2°C)
Chinese cabbage	24–28°F (–4 to –2°C)
Kale	24°F (–4°C)
Leeks	20°F (–7°C)
Brussels sprouts	20°F (–7°C)

Root Cellaring:
Putting Food By the Easy Way

Our garden's focus has shifted across the years. As we became better gardeners, we wanted to grow at a larger scale. We are finally doing just that, and now our focus is self-sufficiency. After we built the garden, the root cellar followed close behind. Root cellars are used to store both root vegetables that need a high moisture environment and those that need low humidity. Lower humidity is achieved quite easily in most northern climates, in drafty homes. We simply use a closet with an exterior wall to store our winter squash, dry corn, garlic, onions, and dry beans.

Our larger and more insulated root cellar is a walk-in closet with ventilation that we built into our garage. This damp room stores all the traditional root cellar vegetables: potatoes, carrots, radishes, cabbage, Brussels sprouts, beets, and leeks. We maintain humidity between 90 and 95 percent all winter, with temperatures between 33 and 39°F (0 and 4°C), aided by damp sawdust in wire baskets where the roots are stored.

The first fall frost, though not a killing freeze, dusted some of the annual flowers, which kept on blooming for several more weeks, albeit ragged and worn.

Our fall garden leans heavily toward what will store well in our root cellars. We look to turn over beds in favor of more midseason succession generalists, such as carrots, beets, or daikon, for winter meals. The beauty of fall gardening for the root cellar is that the food stores fresh. Walking into the root cellar to grab a Chinese cabbage for a January stir fry is nothing short of marvelous. I guarantee this will continue to bring joy and creativity to our garden journey for decades to come.

When I visit the garden after the first fall frost, I'm met with the marvel of ice crystals on my sage and cauliflower leaves, and fiery red blueberry foliage. Rather than address the frost's mess immediately, I enjoy its gifts dotted across the landscape. It's a shift both in mind and hand, anticipating and letting go of summer loves. Autumn also reminds me we have the opportunity, and perhaps even the obligation, to grow right along with the garden every year. It's what makes these waning weeks of warmth the best moment to think back on the growing season and set intentions for the following year. It's the time to dream big.

Bringing all these strategies to the succession garden across the seasons will create a dynamic home garden that will feed you longer, and with more diversity. Every season has a handful of key vegetables instrumental to stretching that season. Let your seasons bump into one another. Tend your soil, renew spent plantings, incorporate perennials into your landscape, and map out how to get ahead in spring and fall—each of these strategies and more can help to extend your growing season. May they also bring you delight in the exploration and realization of what your garden can become.

Late fall in the damp root cellar includes baskets of bok choy, Chinese cabbage, and Brussels sprouts (dug up and repotted for storage), plus buried carrots, daikon, radish, kohlrabi, and potatoes.

OPPOSITE
Succession gardening: a way of life, an act of hope and renewal.

Succession gardening is a lifestyle, to be sure. As with the most hard-won ways of life, its beauty deepens with each passing year as the lessons of the garden further foster our affection and curiosity for what else is possible. It is an act of hope and renewal in ourselves—and the world.

Adventitious roots Roots arising from a place other than the plant's roots, most commonly from a rhizome or underground stem.

Canopy The collective structure that trees (and plants) create in a woodland or forest (or garden), often creating complete cover.

Determinate A plant that grows to a specific maturation, sets fruit in a short window of time, and then dies; paste tomatoes are commonly determinate, which is why they are great for canning.

Direct sowing A common method of planting seeds directly into prepared ground.

Early succession A plant or group of plants that reach maturity in the shortest amount of time relative to other plants around them.

Etiolation The literal stretching of a plant stem in the absence of proper sunlight, a process that weakens the plant often to the point of no return.

Indeterminate The indefinite production of fruit on mature plants; tomatoes, cucumbers, and summer squash are all examples.

Indoor sowing A method of starting seed by planting in pots indoors. Most often implemented to give slow-growing vegetables or flowers a head start before the outdoor growing season begins, so they have plenty of time to mature.

Interplanting A way of planting different plants together for aesthetics, succession planting, or sometimes for mutually beneficial reasons (known as companion planting).

Late succession Plants that take a long time to mature relative to the plants around them and, in the case of food gardening, include fruit trees that take years to bear, as well as alliums and Brussels sprouts, which require half the year in the ground to reach maturity.

Leaf area The area leaves occupy spatially both at a given height and as a whole. The leaf area of an overstory plant determines how much sunlight reaches the understory, and is important to consider when interplanting in the garden.

Low tunnel A temporary garden structure that is erected to protect plants from freezing temperatures, most commonly a half moon shaped structure resembling a caterpillar.

Mid-succession Plants that take part but not all of a season to mature and bear fruit. This includes carrots and many root vegetables, cucurbits like cucumber and summer squash, and when planted as starts, tomatoes, peppers, and eggplants.

Overstory The tallest plants in a planting, or the tallest mature trees in a forest, which occupy the uppermost height of a given space.

Row cover A thick material, plastic or woven fabric, that is secured to the outside of low tunnels as the barrier to protect against freezing temperatures.

Senescence The natural process of mature plants as they transition from living to dying.

Sowing The act of planting of seeds in soil or other medium.

Starts Young plants started indoors to give them extra time to establish before being planted in the garden. This is most common for tomatoes, eggplants, and peppers, which are hot season crops that take a few months before they begin bearing fruit.

Stolon An underground stem that grows laterally (horizontally) to a bud (node) where roots cluster, and a new, identical plant forms. Strawberries and grasses commonly reproduce through stolons, also called runners.

Succession Transitioning from one plant to another in the same space across an entire growing season. In simplest terms, one follows another.

Understory Plants that occupy the lowest portion of a given area, usually in part or full shade situations in a forest. In the food garden, the understory is any low-growing food or flowers that are interplanted with vining crops or taller plants.

Favorite Seed and Fruit Tree Catalog Companies

Adaptive Seeds
Sweet Home, Oregon
Open-pollinated, organic flower and vegetable seed company
adaptiveseeds.com

Botanical Interests
Broomfield, Colorado
A blend of organic and conventional heirloom, open-pollinated, and hybrid flower and vegetable seeds
botanicalinterests.com

Cummins Nursery
Ithaca, NY
Specializes in fruit trees: apples, pears, cherries, peaches, nectarines, and plums
cumminsnursery.com

Fruition Seeds
Naples, New York
Organic, open-pollinated seeds adapted to northern climates
fruitionseeds.com

High Mowing Organic Seeds
Wolcott, Vermont
A blend of heirloom, open-pollinated, and hybrid flower and vegetable varieties
highmowingseeds.com

Honeyberry USA
Bagley, MN
Specializing in cold-hardy fruit shrubs, both cultivated and native varieties
honeyberryusa.com

Johnny's Selected Seeds
Albion, Maine
A blend of organic and conventional, open-pollinated, and hybrid flower and vegetable varieties
johnnyseeds.com

One Green World
Portland, OR
Wide array of unique and rare native and cultivated perennial edibles for all growing zones
onegreenworld.com

Saint Lawrence Nursery
Potsdam, NY
Growing fruit tree stock in chilly zone 3, this is an excellent resource for cold-climate orchard enthusiasts
stlawrencenurseries.com

Seed Savers Exchange
Decorah, Iowa
A nonprofit dedicated to heirloom and open pollinated flower and vegetable varieties, they curate a 20,000+ seed bank and work to fulfill their mission to keep diversity thriving by distributing their seed collections into gardens everywhere
seedsavers.org

Southern Exposure Seed Exchange
Mineral, Virginia
A blend of organic and conventional, open-pollinated and hybrid flower and vegetable varieties, specializing in Mid-Atlantic and Southeast climates
southernexposure.com

Territorial Seed Company
Cottage Grove, Oregon
A blend of organic and conventional, open-pollinated, and hybrid flower and vegetable varieties
territorialseed.com

Western Native Seed
Coaldale, Colorado
Offers seed native to high desert of the interior west, like Colorado's front range, and plants adapted to droughty conditions
westernnativeseed.com

Recommended Reading

Gardening

The New Seed Starters Handbook by Nancy Bubel with Jean Nick

The New Organic Grower: A Master's Manual of Tools and Techniques for the Home and Market Gardener by Eliot Coleman

Four-Season Harvest: Organic Vegetables from Your Home Garden All Year Long by Eliot Coleman

Charles Dowding's No Dig Gardening Course 1: From Weeds to Vegetables Easily and Quickly by Charles Dowding

The Year-Round Vegetable Gardener: How to Grow Your Own Food 365 Days a Year, No Matter Where You Live by Niki Jabbour

Weedless Gardening by Lee Reich

Growing Vegetables West of the Cascades: The Complete Guide to Organic Gardening by Steve Solomon

Reference

Garden Insects of North America: The Ultimate Guide to Backyard Bugs by Whitney Cranshaw and David Shetlar

Bees: An Identification and Native Plant Forage Guide by Heather Holm

Wasps: Their Biology, Diversity, and Role as Beneficial Insects and Pollinators of Native Plants by Heather Holm

The Organic Gardener's Handbook of Natural Insect and Disease Control Edited by Barbara W. Ellis, Fern Marshall Bradley, and Deborah L. Martin

Good Bug Bad Bug: Who's Who, What They Do, and How to Manage Them Organically by Jessica Walliser

Attracting Beneficial Bugs to Your Garden: A Natural Approach to Pest Control by Jessica Walliser

ACKNOWLEDGMENTS

Writing a book has long been a goal of mine. And it would not have been possible without the support of numerous friends, colleagues, and my loving family.

To my acquiring editor, Tom Fischer, project editor, Michael Dempsey, and the entire Timber Press team who championed this from the very first seed I shared. Your belief in this book and my abilities have been just the fuel I needed to persevere. And a special thanks to Andy Keys Pepper for diving into the weeds with me, sharpening the book's focus, and doing so in the most supportive way. You all made this project fun. Truly.

To all of my sounding boards and readers who helped me unearth more clarity along the way. Thank you especially Michael Larsen, Anna White, Mia Bolte, and Doug Cowden. You each added much-appreciated perspectives, caught important edits, and helped add ideas that propelled this book to become even better. Thank you for your time, insight, encouragement, and thoughtful feedback. I love each of you dearly.

To my sweet friend Ashlie Blake who created the most beautiful illustrations. Thank you for bringing your talents to the pages of this book.

To my friends Dan and Kathleen Walsh, whose quiet getaway helped accelerate the writing process. Thank you for sharing your slice of driftless Wisconsin with me.

To my friends who opened their land to me to help round out the photographs for this book. Dean Engelmann of Wise Acre Farm, thank you for welcoming a near stranger to your fields, graciously inviting me to wander and capture the magic that is an early August evening on your organic farm. Rachel Henderson and Anton Ptak of Mary Dirty Face Farm, thanks for allowing me to capture the beauty of a mature fruit orchard. Kathy Thomas and John Ruggles, both old friends and fellow garden enthusiasts, welcomed me on a whim to capture some early evening garden glow in a beautifully planted multi-generational family garden. Jake Lau and Briana Odegard of Square Roots Farm in Fall Creek, Wisconsin, thank you for welcoming me on a random sunny early summer day to explore your beautifully productive farm and capture your cover crops.

To my wonderful network of authors and garden friends who lent an ear in the early phase of my process: Brie Arthur, Kris Bordessa, Nicole

Burke, Kevin Espiritu, Niki Jabbour, Margaret Roach, Anne Marie Ruff, Mary Schier, and Lisa Steele. Thank you all for sharing your book publishing experiences with me.

A special thanks to Steve Solomon whose email communications over the course of my book writing were wonderfully supportive—and helped me update an old organic fertilizer recipe.

And special thank yous to Emily Murphy and Joe Lamp'l for your continued support throughout the process. Your mentorship throughout this process has been invaluable. Thank you both for your friendship and support.

Thanks to all my friends, old and new, near and far, online and in real life, who cheered me on throughout this process. Your belief in me helped me tremendously. To my family, near and far, who knew me when a book was a hazy but nagging vision I couldn't shake—thank you. A special thank you to my sister, Beth, who was a steadfast supporter from start to finish.

To my husband and best friend, John, and our children, who have always supported and believed in me, even when I procrastinated for weeks on end. Our annual experiments and dedication to constantly growing and learning together, in the garden and in life, have brought richness to my life in ways that continue to delight me—and for that I am eternally grateful. I love you.

A

Achillea millefolium, 134
African marigolds, 208
Alaska nasturtium, 215, 238
All-America Selections (AAS), 204
alliums, 49, 116–117
AmaRosa potatoes, 35, 114, 115
Amelanchier alnifolia, 67
Amelanchier spp., 75
American cranberrybush, 67
American hazelnut, 67
American plum, 74, 75
Amish paste tomatoes, 108
anise hyssop, 134
Anna apples, 72
annual/cut flower sowing guide, 186
anthracnose fungal disease, 258
aphids, 236
apples, 27, 54, 56, 72, 73, 74, 76, 78, 80, 81, 136
apricots, 72, 76, 81
Arbor Day Foundation, 74
Armenian cucumbers, 103, 104
arugula, 40, 44, 87, 90, 95, 99, 172, 185, 194, 196, 228, 232, 248, 251, 253, 254
Asclepias spp., 134
Asimina triloba, 75
asparagus, 68, 69, 81, 82, 84, 256, 267
aspens, 50
aster, 134
auxiliary gardens, 132–136, 215, 216, 255
avocados, 72
Aztec marigolds, 208

B

Back Cherry tomatoes, 108
back-to-back planting, 34–35
bareroot plants, 73, 76, 267, 268
basil, 124, 126, 184, 194, 219–220, 241, 246, 248, 250
beans, 17, 31, 34, 46, 48, 52, 98, 99–101, 228, 239, 248, 253, 255, 258, 265
beardtongue, 134
Beas kohlrabi, 166
bed maintenance, 190, 249
beefsteak tomatoes, 31, 108, 110
bees, 42, 50, 131, 132–134, 138, 139, 141, 206
beets, 29, 44, 87, 111, 120, 159, 172, 177, 183, 184, 196, 228, 232, 233, 246, 247, 248, 250, 253, 254, 258, 259, 264, 270
bell peppers, 31, 110, 233
Benary's Giant zinnia, 198, 203
berries
    blackberries, 61–62
    blueberries, 83, 190, 255, 267, 268, 272
    compost, 190
    cranberry, 67
    elderberry, 56, 67, 76, 255
    gooseberry, 67
    honeyberry, 65
    lingonberry, 67
    mulch, 190
    native edible shrubs, 66–67
    planting, 81–83
    raspberries, 61–62, 76, 236, 256
    Saskatoon berry, 67
strawberries, 58–61, 73, 76, 78, 249, 267, 268
    types, 56–67
birds, 34, 42
blackberries, 61–62
black-eyed Susan, 132, 140
black Spanish radish, 251
blocking, 26, 33–35, 44
Bloody Mary nasturtium, 215
blueberries, 61, 62–64, 83, 190, 255, 267, 268, 272
blue vervain, 134
Blue Wind broccoli, 166
bok choy, 40, 44, 90, 99, 156, 172, 183, 184, 233, 248, 250, 253, 254, 258, 259, 270, 272
Boothby Blonde cucumbers, 103
boron, 52
Brandywine tomatoes, 108
Brassica oleracea, 88, 94, 118
brassicas, 21, 34, 38, 43, 49, 70, 88–97, 99, 124, 159, 178, 181, 187, 196, 225, 232, 239, 245, 253, 254, 255, 265, 268
breadseed poppy, 186, 217
Bright Lights cosmos, 205
broccoli, 31, 34, 38, 50, 87, 88, 90, 95, 96, 118, 159, 166, 172, 183, 184, 225, 228, 232, 233, 246, 247, 248, 250, 253, 254, 259, 265, 270
broccoli raab, 95, 172, 184, 248, 251, 253, 254
Brussels sprouts, 27, 40, 87, 88, 95, 118–119, 183, 184, 194, 196, 227, 228, 233, 250, 263, 269, 270, 272
buckwheat, 34, 153, 236, 254
Buddha's Hand cosmos, 205
bumblebees, 133, 134
burpless cucumbers, 103
bush beans, 32, 48, 98, 100, 101, 185, 194, 241, 251, 253, 254, 261, 264, 265
Butte potatoes, 115

butterflies, 128, 132, 204
butternut squash, 101

C

cabbage, 34, 38, 87, 88, 90, 91, 95, 97, 98, 111, 118, 159, 166, 172, 183, 184, 187, 192, 203, 225, 228, 233, 238, 246, 247, 248, 250, 253, 254, 259, 265, 269, 270
cabbage moths, 37
calendula, 50, 186, 194, 201–203, 212, 218, 228, 238, 239, 240, 259, 270
Canada wild rye, 15
canes, 61
canopy diversity, 17–18, 22–23, 50, 226
Caribé potatoes, 115
Carola potatoes, 115
carrots, 21, 22, 34, 44, 87, 111, 121–123, 183, 185, 192, 194, 196, 228, 233, 241, 246, 247, 248, 249, 251, 253, 254, 259, 263, 264, 269, 270, 272
cauliflower, 52, 88, 90, 95, 96, 118, 172, 183, 184, 192, 225, 228, 232, 233, 246, 248, 250, 254, 270
cedar waxwings, 68
celery, 34, 38, 172, 194, 246, 248, 250, 254
chamomile, 236
Chantenay carrots, 123
Charlotte potatoes, 115
Cherokee Purple tomatoes, 108
cherries, 72, 74
cherry tomatoes, 27, 31, 106, 108, 110, 260
Chinese cabbage, 40, 172, 183, 184, 248, 250, 254, 259, 270, 272
chives, 68, 126, 220
chokeberry, 67
cilantro, 87, 126, 184, 194, 196, 248, 250, 251, 259

climate change, 142, 177
clover, 137, 139, 199
cold tolerance, 119, 214, 268, 270
collards, 118
common elderberry, 68
common persimmon, 75
common yarrow, 134
companion flowers, 213–219
composites
    annual flower successions, 212
    calendula, 50, 186, 194, 201–203
    cosmos, 186, 194, 205–206
    marigolds, 186, 194, 207–208
    Mexican sunflower, 186, 209
    sunflowers, 186, 194, 210–211
    types, 201–211
    zinnia, 52, 186, 194, 203–204
compost, 145, 146, 149, 190, 256, 267
coneflower, 15, 140
continuous planting, 26, 27–29
corn, 10, 39, 47–48, 87, 102, 112–113, 183, 185, 194, 231, 233, 248, 249, 256, 265
Corylus americana, 67
cosmos
    interplanting, 206, 228, 229
    plant spacing, 218
    sowing, 186, 194
    succession of, 212
    types, 205
Cosmos bipinnatus
    Cupcake, 205
    Double Click, 205
    Rubinato, 205
    Seashells, 205
    Sensation, 205
    Versailles, 205
Cosmos sulphureus
    Bright Lights, 205
    Buddha's Hand, 205
    Sulphur, 205
cover crops, 33, 34, 153, 254
crabapples, 80

Crackerjack marigold, 208
crop rotation, 33, 149, 267
cross-pollination, 80, 140
cucamelon, 103, 184, 231
cuckoo bees, 133
cucumbers, 46, 87, 101, 102–104, 183, 185, 192, 194, 196, 227, 228, 229, 233, 239, 241, 246, 248, 249, 250, 252, 254, 258, 259, 261, 266
cucurbits, 46, 48, 106, 196
cultivars, 58, 62, 66, 73, 96, 139–140, 211, 219
Cupcake cosmos, 205

D

dahlias, 235
daikon, 248, 251, 254, 259, 264, 272
Danvers 126 carrots, 123
Danvers carrots, 123
Dapple Grey dry beans, 101
day-neutral strawberries, 59
deer, 69, 77, 106
deer-proof fence, 77, 78, 106
deer-resistant plants, 70, 106, 220
dent corn, 47, 112, 113, 185, 196
determinate plants, 31, 100, 108, 254
dill, 126, 194, 221, 246, 248, 250, 259
Diospyros virginiana, 75
direct sowing, 29, 37, 85, 99, 120, 175, 177, 182, 190, 194, 196, 211, 246, 249
disease, 52, 245, 258
diversity, 19–20
dominant plantings, 227
Double Click cosmos, 205
Double Gold raspberries, 63
dry beans, 100, 185, 194, 246, 248, 249
Durango marigold, 208
Dutch white clover, 139
dwarf fruit trees, 78
dwarf root stock, 73

E

early bearing strawberries, 73
early figwort, 132
early fruits, 56–67
early harvests, 172
early succession vegetables, 94–97
echinacea, 140
edamame, 100, 185, 251, 253, 256
edges, 20–21, 50, 51, 77, 81, 218, 204, 226, 235, 239, 249
eggplant, 38, 50, 68, 78, 87, 99, 109–110, 159, 172, 183, 196, 203, 238, 239
elderberry, 56, 67, 76, 255
Empress of India nasturtium, 215
endurance, 26, 36–38
English cucumbers, 103
espalier dwarf trees, 73
espalier fruit trees, 77–79, 81
espalier tomatoes, 107
etiolation, 165
Eutrochium purpureum, 134
everbearing strawberries, 59, 73

F

fall gardening
    fall garden tasks, 267
    fall harvests, 250–251
    planning, 10, 264–269
fennel, 126, 183, 184, 194, 248, 250
field corn, 113
Fiesta nasturtium, 215
figs, 72
filet beans, 100
first freeze, 268–269
first frost, 269, 272
flea beetles, 37
flint corn, 47, 112, 113, 185, 194
floating row covers, 255–256
floricanes, 61

flour corn, 228
flowers
    annual/cut flower sowing guide, 186
    annual flower successions, 198, 199–221
    auxiliary gardens, 132–136, 215, 216, 255
    bedding, 161
    berries, 56, 65
    companions, 213–219
    composites, 201–211, 212
    cover crops, 153
    edges for, 20
    as foundation plants, 236–237
    generalists, 238
    indoor sowing, 180
    insectaries, 236
    interplanting, 38, 41–46, 50–53, 78, 86, 106, 132, 199–218, 222, 228, 234–238
    late succession, 235
    lawns, 137, 139, 146
    location, 235
    May sowings, 194
    native perennials, 134
    native plants, 139–143
    organic fertilizer, 150
    planning, 29
    plant spacing, 218
    prairies, 14–15, 21, 216
    repetition, 236
    starts, 159, 173
    summer gardening, 246
    vegetable combinations, 228
    wildflowers, 141–143
fluorescent bulbs, 168, 169
forests, 13, 17–23, 27, 30
foundation plants, 236–237
French marigolds, 208
frost, 16, 27, 36, 38, 40, 85, 89, 91, 96, 97, 99, 102, 175, 196, 248, 258, 264, 269, 272

frost tolerant plants, 89, 91, 96–97, 269
fruit trees, 70–81, 239, 245, 249, 268
Fuji apples, 72

G

garbanzo beans, 17, 52, 100
garden space, 157
garlic, 29, 34, 40, 87, 106, 116–117, 233, 246, 253, 254, 268
garlic scapes, 117
generalist succession vegetables, 87, 119–126, 196
Genovese basil, 124, 220
German Butterball potatoes, 115
germination
    asparagus, 82
    beets, 120
    carrots, 121
    indoor seed starting, 164, 165, 166, 168, 177, 187
    mats, 164, 168
    soil temperatures, 192
Giant orange marigold, 208
Giant yellow marigold, 208
ginger, 159
globe amaranth, 186, 216
golden Alexanders, 134
goldenrod, 15, 134
Gold Rush bush beans, 101
Goldy summer squash, 105
gooseberry, 67, 172
Granny Smith apples, 72
grapefruit, 72
grape tomatoes, 108
green beans, 29, 34, 87, 183, 233, 246, 247, 248, 259, 261, 266
green onions, 40, 172, 259
ground cherries, 110, 184
ground-nesting bees, 131
ground nuts, 35

H

Haralson apples, 72, 136
hardening off, 173–175
hardneck garlic, 117
harvest season, 13, 157
hazelnut, 66, 77, 83
head lettuce, 29, 87, 159, 184, 194, 225, 233, 248, 250, 254, 258, 259, 270
heat-tolerant plants, 29, 90, 96–97, 194, 196, 246, 247, 250, 254, 259, 265
heirloom tomatoes, 108
herbs, 49, 78, 87, 124, 126, 194, 196, 219–221, 228, 236, 239, 246, 259
highbush blueberry, 62
highbush cranberry, 67
holy basil, 219
honeybees, 133, 138, 206
honeyberry, 65
Honeycrisp apples, 72
hot peppers, 31, 110, 233
hoverflies, 239
hummingbirds, 236
hybrid plants, 92, 105, 166, 204, 211

I

Imperator carrots, 123
indeterminate plants, 98, 100, 108, 227, 229, 235, 241, 269
indoor sowing, 37–38, 85, 99, 120, 159, 164–171, 194, 221, 246
industrial farming, 33–34
in-ground beds, 35, 146
insectaries, 236
insects, 12, 34, 41, 42, 43, 46, 50–51, 131, 132–136
interplanting
    art of, 223–241
    beans, 101
    berries, 63
    best practices, 224–230
    blocking and, 34–35

complementary foliage, 229
dominant plantings, 227
edges, 20, 50, 51, 81, 204, 218, 226, 235, 239, 249
evenly growing plants, 225
flowers, 38, 41–46, 50–53, 78, 86, 106, 132, 199–218, 222, 228, 234–238
fruit trees, 78, 80
gaps, 225
greens, 90, 91
herbs, 219–221
indoor sowing, 159
low-growing plants, 48–51, 52, 196, 238, 239
overplanting and, 51–53, 232
plant structure/form considerations, 230
role of sunlight, 226
small space strategies, 241
staggered maturities, 43–44, 229
as strategy, 41–53
summer gardening, 255
summer squash, 106
three sisters, 48
vegetable combinations, 228
vertical gardening, 48, 231

J

Japanese beetle, 142
Japanese beetles, 249, 256
Jazzy Mix zinnia, 204
Jewel nasturtium, 215
June-bearing strawberries, 59, 73
June sowings, 246

K

kale, 21, 87, 88, 90, 91, 118, 172, 183, 184, 194, 225, 233, 246, 248, 250, 259, 263, 269, 270
Keera cucumbers, 103
Kennebec potatoes, 115

kimchi, 40
kohlrabi, 18, 37, 40, 44, 87, 88, 91–92, 95, 118, 166, 172, 181, 183, 184, 196, 233, 248, 250, 253, 254, 258, 259, 270, 272
Kolibri kohlrabi, 166

L

lacewings, 236
Ladybird Rose nasturtium, 215
late succession flowers, 235
late succession vegetables, 87, 94–97, 112–119, 196
lavender, 126
lawns, 137–139, 146
leaf area, 18, 223, 227, 229
leafcutter bees, 133
leaf lettuce, 44, 90, 185, 232, 251, 253
leaf mulch, 145
leafy greens, 87, 88–91
LED bulbs, 168, 169
leeks, 87, 116–117, 172, 183, 196, 263, 270
lemon cucumbers, 103, 104
Lemon Drop marigold, 208
Lemon Gem marigold, 208
lemons, 72
lemon verbena, 126
lettuce, 29, 34, 40, 41, 49, 50, 172, 177, 183, 192, 194, 196, 228, 241, 246, 247, 248, 250, 254
light availability, 17, 226
lingonberry, 67
location, 235
low-growing plants, 48–51, 52, 196, 238, 239
low tunnel enclosures, 39–40, 159, 160, 180–181, 187, 190, 191, 266

M

Magic Molly potatoes, 115
mandarin, 72

marigolds
  interplanting, 228, 234, 238, 239, 240
  plant spacing, 218
  sowing, 186, 194
  succession of, 212
  types, 207–208
Marketmore76 cucumbers, 103, 104
mason bees, 133
Maxibel bush beans, 101
May sowings, 194
meadow blazing star, 128
melons, 46, 183, 185, 194, 246, 249, 258
Mexican sunflower, 186, 209, 218
micro-monoculture planting, 35
mid-bearing strawberries, 73
midseason succession flowers, 235
midseason succession vegetables, 87, 98–111, 196, 197
mid-succession fruits, 56–67
Miles garbanzo beans, 101
milkweed, 134
mindfulness, 29
mint, 126
mixed species forests, 22–23
mizuna, 254
monarch butterflies, 42, 128, 132
Monarda fistulosa, 134
monoculture planting, 22, 33–34
Morus rubra, 75
mountain mint, 134
mulberry, 74
mulch, 17, 145, 149, 190
mustard greens, 87, 90, 184, 248, 251, 254, 270

N

Nantes carrots, 123
nasturtiums, 186, 194, 215, 218, 228, 230, 238, 239, 240
Native American agriculture, 48

native bees, 131, 133–134, 139, 141
native edible shrubs, 66–67
native flowers, 139–143
native trees, 74–76
nectarines, 72
nematodes, 256
newspaper pots, 187, 188–189, 196
nightshade family, 68–69
Norland potatoes, 115
Northern Lights zinnia, 204
no-till organic gardening, 146–147
novelty cucumbers, 103
nut trees, 70

O

oak, 21, 66
oak savannas, 21
Ohio spiderwort, 14
Oklahoma zinnia, 204
okra, 185
onions, 18, 38, 40, 50, 87, 106, 116–117, 159, 172, 178, 183, 192, 228, 233, 253, 261, 266, 267
Opal basil, 124, 220
open-pollinated cucumbers, 103
open-pollinated tomatoes, 108
orangerie, 39
oranges, 39
Orchid Cream nasturtium, 215
oregano, 126
organic fertilizer, 145, 150–151
overplanting, 51–53, 232
overstory, 14, 17, 19, 20, 21, 76, 227, 228
oxheart carrots, 123

P

pale purple coneflower, 140
Papaver somniferum cvs., 217
parsley, 126, 184, 250
paste tomatoes, 31, 108, 156

paw paw, 75
peaches, 72, 266
peanuts, 35
pears, 56, 70, 74, 76, 245
peas, 33, 34, 87, 111, 225, 248, 259, 270
pecans, 74
Penstemon spp., 134
peppers, 27, 31–32, 38, 50, 68, 78, 87, 99, 109–110, 159, 172, 178, 183, 184, 192, 196, 228, 238, 239, 261, 264, 266
perennial fruit succession, 64
perennial vegetables, 68–70
Persian Carpet zinnia, 204
Persian cucumbers, 103, 104
pesto, 220
pests, 36, 37, 83, 249, 255–258
Petite marigold, 208
pickling cucumbers, 103, 104, 156, 254
Pink Lady apples, 72
pinto beans, 100
planning, 25, 29
plant spacing, 17, 20–21, 51–53, 111, 218, 232, 233
plant stress, 52
"plant the garden" holidays, 99
plum curculio, 249
plums, 56, 72, 74, 76, 81
plum tomatoes, 108
Polar Bear zinnia, 204
pole beans, 32, 100, 101, 185, 194, 231, 256
pollinators, 20, 42, 44, 46, 80, 129, 132–134, 138, 139, 142, 204, 210, 213–214
popping corn, 112, 113, 185, 196, 228
potatoes, 35, 40, 49, 68, 87, 114–115, 183, 185, 194, 196, 228, 233, 251, 254, 255, 272
potting up, 159
powdery mildew, 52

prairies, 13–17, 27, 30, 41–42, 131, 143, 240
primocanes, 61
pruning, 61, 104, 106, 159
Prunus americana, 75
purple meadow rue, 14
Pycnanthemum virgianum, 134

Q

Queen Lime zinnia, 204
quick succession vegetables, 87, 88–94, 196

R

rabbit-proof fence, 77, 78, 255
rabbits, 77, 255
radicchio, 172, 184, 248, 250
radishes, 34, 40, 44, 52, 87, 88, 89, 91, 95, 98, 99, 172, 181, 183, 185, 191, 192, 194, 196, 225, 228, 232, 233, 248, 251, 253, 254, 270, 272
raised beds, 20
raspberries, 61–62, 76, 236, 256
red cabbage, 41, 44, 49, 97, 227, 232, 254
Red Gem marigold, 208
red mulberry, 75
Red Pontiac potatoes, 115
repetition, 33, 236
repetition planting, 33
Resina calendula, 202
rhubarb, 68, 69–70, 106
Ribes spp., 67
romaine lettuce, 49
Romanesco broccoli, 44, 95, 172, 184, 187, 225, 250
Romano beans, 100
root cellars, 17, 271–272
Rose Finn potatoes, 115
rosemary, 126
Rosie O'Day sweet alyssum, 240

row covers, 17, 29, 180, 255–256, 267
Rubinato cosmos, 205
runners, 57, 59, 60, 62
Russet potatoes, 115
Russian Banana potatoes, 115
Russian Mammoth sunflower, 210
rutabaga, 183, 248, 251, 259, 270

S

sage, 126, 272
salad gardens, 20, 29
Sambucus spp., 67
Saskatoon berry, 67
sauce tomatoes, 108
savoy cabbage, 44, 49
scarlet kale, 22
sea kale, 70
Seashells cosmos, 205
seedless cucumbers, 103
seed starting, 29, 36–38, 82, 85, 126, 159, 160, 161–177, 190, 249
seed starting hardware, 168, 169
self-seeding flowers, 141
semi-dwarf trees, 73
senescence, 245
Senora zinnia, 204
Sensation cosmos, 205
serviceberry, 75
shallots, 116–117, 172, 183
shelling beans, 100
shoulder season plants, 38, 95, 98–99, 196–197, 201, 216
shrubs, 81
signet marigolds
  Lemon Gem, 208
  Red Gem, 208
  Tangerine Gem, 207, 208
single crop planting, 30–32
single successions, 33
slicing cucumbers, 103, 104
slicing tomatoes, 108

small space strategies, 241
snap beans, 192, 196, 248, 265
snapdragon, 50, 186, 212, 216, 228, 270
snap peas, 34, 183, 185, 233, 251
sneezeweed, 134
snow peas, 34, 183, 185, 233, 251
soil
  feeding, 149–153, 256
  health, 145, 148, 153
  no-till organic gardening, 146–147
  pH, 148
  temperatures, 87, 99, 164, 177, 181, 182, 190, 192–193, 196
  tending, 145–147
  testing, 148, 267
soil blocking, 166, 170
solanaceous crops, 99, 106–110
soldier beetle, 142
Solidago spp., 134
Sophia marigold, 208
sorrel, 68
sowing, 29, 37–38, 85, 177, 180–187, 190, 194, 196, 211, 246, 249. *See also* direct sowing, indoor sowing
soybeans, 100
Speckled Cranberry dry beans, 101
spinach, 40, 44, 87, 90, 91, 172, 183, 185, 192, 196, 232, 248, 251, 253, 254
spring gardening
  early sowing, 180–187
  early spring tasks, 190
  layering successions, 196–197
  May sowings, 194
  radishes, 191, 225
  soil temperatures for germination, 192–193
  succession planting guide for continuous harvests, 184–185
sprouting broccoli, 17

squash, 44, 46, 48, 52, 120, 132, 228, 239, 255, 260
staggered maturities, 26, 30–32, 43–44
starts, 37, 48, 90, 97, 101, 110, 117, 126, 141, 159, 161–164, 168, 169, 173–177, 180, 182, 190, 191, 249, 253, 254, 258, 264
State Fair zinnia, 204
stolons, 61
strawberries, 58–61, 73, 76, 78, 249, 267, 268
strawberry renovation, 60
strawflower, 186, 217
string beans, 99, 194, 249
succession, 12–13
succession gardening
    bed maintenance, 190, 249
    blocking, 26, 33–35
    continuous planting, 26, 27–29
    early spring tasks, 190
    early succession vegetables, 94–97
    endurance, 26, 36–38
    fall gardening, 263–273
    flowers with food, 26, 45–46
    garden layout, 253–254
    garden space and harvest by season, 183
    generalist succession vegetables, 87, 119–126, 196
    harvest season, 13
    interplanting, 20, 26, 34–35, 38, 41–53, 78, 80, 86, 90, 91, 106, 159, 222, 223–242, 255
    late succession vegetables, 87, 94–97, 112–119, 196
    layering successions, 196–197
    low-growing plants, 48–51, 52, 196, 238, 239,
    management, 16–17, 35, 61
    midseason succession vegetables, 87, 98–111, 196, 197
    overplanting, 51–53, 232

perennial fruit succession, 64
planning, 25, 29, 37, 126, 156–159, 179, 180, 196–197, 226, 241
plant spacing, 17, 20–21, 51–53, 111, 218, 232, 233
quick succession vegetables, 87, 88–94, 196
shoulder season plants, 38, 95, 98–99, 196–197, 201, 216
spring gardening, 179–197
staggered maturities, 26, 30–32, 43–44
strategies for, 26–53
successions within successions, 73
summer gardening, 243–262
types, 87–121
variety, 26, 30–32
vertical gardening, 26, 46–48, 101, 102–104, 116, 231, 241
winter succession planting guide, 172
winter tasks, 160
zone bending, 26, 38–40, 175
Sugarsnax carrots, 123
Sulphur cosmos, 205
summer gardening
    June sowings, 246
    late summer gardens, 260
    layering successions, 261–262
    pests, 255–258
    planning, 249–255
    planting guide for extended fall harvests, 250–251
    sowings for fall garden, 259
    succession planting, 243–262
    summer garden tasks, 249
summer squash, 44, 84, 87, 105–106, 183, 185, 194, 196, 228, 233, 246, 248, 249, 250, 253, 254, 257, 261, 266
sunflowers, 186, 194, 210–211, 212, 218, 228, 231, 235, 236, 246, 260

sunlight, 17, 226
Suyo Long cucumbers, 103
sweat bees, 133, 140
sweet alyssum, 41, 50, 52, 186, 194, 212, 213–215, 218, 228, 238, 239, 240, 259, 270
sweet corn, 39, 102, 112, 185, 192, 194, 196, 246, 251, 252, 253, 254, 265
sweet Joe-Pye weed, 134
Symphyotrichum spp., 134

T

Tagetes erecta
    Crackerjack, 208
    Giant orange, 208
    Giant yellow, 208
Tagetes patula
    Durango, 208
    Lemon Drop, 208
    Petite, 208
    Sophia, 208
Tagetes tenuifolia, 208
Tall Trailing nasturtium, 215
Tangerine Gem marigold, 207, 208
Terek kohlrabi, 166
Thai basil, 124
thermometers, 182
three sisters, 48
Thumbelina zinnia, 204
thyme, 126
Tiara cabbage, 166
Tiger's Eye dry beans, 101
tomatillos, 110, 184, 230
tomatoes, 18, 30–31, 38, 46, 68, 87, 99, 106–109, 132, 159, 172, 177, 178, 183, 184, 196, 228, 229, 233, 239, 241, 254, 255, 260, 264, 265, 266
Toyha edamame, 101
transplanting, 37, 99, 182, 190, 265
trap crops, 255
tree collards, 70

trellises, 46–47, 101, 107, 227, 239, 241
tubers, 35
tulsi basil, 124, 219
turnips, 120, 183, 248, 251, 264, 270

U

understory, 14, 17, 18–19, 20, 21, 22, 76, 213, 215, 227, 228, 229, 230, 235

V

Vaccinum corymbosum, 62
Vaccinum vitis-idaea var. *minimus*, 67
variety, 26, 30–32
Velour bush beans, 101
Verbena hastata, 134
Versailles cosmos, 205
vertical gardening, 26, 46–48, 101, 102–104, 116, 231, 241
viburnum, 255
Viburnum opulus var. *americanum*, 67
viceroy caterpillars, 136
vining plants, 46–47, 50101, 102–104, 106, 108, 208, 218, 229, 231, 241, 269
viola, 270
Violetta cauliflower, 44

W

walnuts, 74
Waltham butternut squash, 116
warm season crops, 190
watermelon, 46, 78, 192
watermelon radish, 248, 251, 254, 259, 264
wax beans, 100

weeds, 35, 41, 51–52, 75, 137, 215, 249, 260
Whirligig Mix zinnia, 204
white false indigo, 14
wild bergamot, 134
wildflower mixes, 141
wildflowers, 141–143
wildlife, 34, 42, 132
wild plum, 74
willow, 136
winter harvesting, 17
winter squash, 40, 48, 87, 116, 183, 185, 194, 196, 233, 246, 269
winter succession planting guide, 172
winter tasks, 160

X

Xerochrysum bracteatum, 217

Y

Yellow Pear tomatoes, 108
Yukon Gold potatoes, 115, 257

Z

Zestar apples, 72
zinnia
    as foundation plants, 237
    interplanting, 228, 235
    overplanting, 52
    planting, 266
    plant spacing, 218
    sowing, 186, 194
    succession of, 212
    types, 203–204
Zinnia elegans
    Benary's Giant, 204
    Northern Lights, 204
    Oklahoma, 204
    Polar Bear, 204
    Queen Lime, 204

Senora, 204
State Fair, 204
Thumbelina, 204
Whirligig Mix, 204
Zinnia haageana
    Jazzy Mix, 204
    Persian Carpet, 204
Zizia spp., 134
zone bending, 26, 38–40, 175
zucchini, 44

Meg Cowden is a blogger and self-taught organic gardener with an advanced degree in natural resource management. Having lived on both coasts and settled in the upper Midwest, she has knowledge and experience living, playing, and gardening in many different ecosystems, from the eastern hardwoods of southern New England where she ran free as a child to the majestic rain forests of the Pacific Northwest where she found home in college, to the edge of the prairie outside Minneapolis where she and her husband are raising their two boys. She has a passion for creating edible landscapes that serve as formal gardens and gathering spaces for more than just humans.